GLACIER EVOLUTION IN A CHANGING WORLD

Edited by **Danilo Godone**

Glacier Evolution in a Changing World
http://dx.doi.org/10.5772/65820
Edited by Danilo Godone

Contributors

Sigmund Hågvar, Mikael Ohlson, Daniel Flø, Cristina Cid, Eva Garcia-Lopez, Rafael Ribeiro, Edson Ramirez, Jefferson Cardia Simões, Yanan Li, Yingkui Li, Xiaoyu Lu, Emil Gachev, Thomas Fickert, Żaneta Polkowska, Sara Lehmann-Koneraa, Marek Ruman, Grzegorz Gajek, Krystyna Kozioła

Published by InTech
Janeza Trdine 9, 51000 Rijeka, Croatia

Publishing Process Manager Martina Usljebrka
Technical Editor SPi Global
Cover InTech Design team

First published October, 2017
Printed in Croatia
Legal deposit, Croatia: National and University Library in Zagreb

Additional hard copies can be obtained from orders@intechopen.com

Glacier Evolution in a Changing World, Edited by Danilo Godone
p. cm.
Print ISBN 978-953-51-3543-2
Online ISBN 978-953-51-3544-9

For EU product safety concerns:
IN TECH d.o.o., Prolaz Marije Krucifikse Kozulić 3, 51000 Rijeka, Croatia,
info@intechopen.com or visit our website at intechopen.com.

Meet the editor

Danilo Godone holds a PhD degree in "Agriculture, Forest, and Food Sciences"; his doctorate topic was cryosphere's phenomena monitoring by geomatic methodologies. Currently, he is a Postdoc grant holder, at Geohazard Monitoring Group (CNR IRPI), studying geomatic contribution in natural hazard monitoring and analysis. His main research interests are landslides, glaciers, and, more generally, natural disasters. During his activities, he has developed skills in GIS, also by developing customized tools by R programming, and land surveying with GPS or laser scanners. He is a member of NATRISK-Research Centre on Natural Risks in Mountain and Hilly Environments, in Turin University. He acts as a freelance consultant, in the same topics, for other research bodies, training agencies, and professionals, too.

Contents

Preface

Glaciers have always played an important role in human history and have been perceived, in ancient times, like monstrous entities of the mountains and are now carefully observed as climate change sentinels. Glaciers influence natural and anthropic systems, e.g., by providing water, influencing local weather, and hosting communities on their body and their surroundings, too. The current climate change scenario is heavily affecting glacier dynamics. Glacier melt rate is rapidly increasing, and without an adequate amount of snowfall, its mass balance is continuously negative. This issue deserves accurate and in-depth studies in order to adequately monitor glacier state and try to mitigate its impact. This circumstance in fact endangers the water supply, thus influencing human settlements, but also creates new environments allowing the colonization by pioneer communities, i.e., primary succession and the formation of new landscapes.

This book is subdivided into two main sections in order to deal with the two topics of worldwide research on glaciers and ecology in glacial environments. In the first one "Glaciers in the World," several reviews and studies are collected. It is an overview of glaciers, their state, and research carried out in different continents and contexts. Chapter 1 concentrates on Svalbard archipelago and deals with cryospheric metrics and analytical chemistry in the environmental monitoring. Chapter 2 focuses on glacier changes in the arid environment of central Asia by applying statistical modeling. The Amazon watershed is analyzed in Chapter 3. It copes with several aspects of this peculiar sector of the cryosphere, i.e., tropical glaciers. At last, Chapter 4 explores the barely known topic of glaciers in Balkan Peninsula with an overview of different mountain ranges and their glacial bodies.

The second section "Glacial Ecosystems" focuses, on the other hand, on glacier environments and ecological researches. Chapter 5 describes glaciers as unique biomes dominated by microbial communities and deals with their role, during glacial retreat, in soil formation and plant settlements. The study of vegetation dynamics, by permanent plots and chronosequences, is addressed in Chapter 6. Chapter 7, finally, deals with 200 years of arthropod succession in a glacier foreland.

I would like to thank all the contributors of this book including the authors of the accepted chapters. My special thanks go to the Publishing Process Manager, Ms. Martina Usljebrka, and the staff of InTech publishing for their kind support. I also thank my colleague, Ms. Martina Cignetti, for providing helpful comments while reviewing this preface.

Danilo Godone, PhD National Research Council,
Research Institute for Hydrogeological Prevention and Protection,
Geohazard Monitoring Group,
Torino, Italy

Section 1

Glaciers in the World

Chapter 1

Glaciers as an Important Element of the World Glacier Monitoring Implemented in Svalbard

Sara Lehmann-Konera, Marek Ruman,
Krystyna Kozioł, Grzegorz Gajek and
Żaneta Polkowska

Additional information is available at the end of the chapter

http://dx.doi.org/10.5772/intechopen.69237

Abstract

Glaciers are not only contributors to the sea level rise but also important players in the circulation of pollutants. Over a billion people apply glacial waters for domestic purposes; hence, both the quality and quantity of this water should be monitored. In this chapter, we concentrate on the archipelago Svalbard in the Arctic, a typical target area for xenobiotics from long range atmospheric transport (LRAT), holding an important share of the Arctic glacial ice cover. Literature review has been conducted over both the cryospheric metrics and the achievements of analytical chemistry in the environmental monitoring. Svalbard is a relatively well-monitored part of the Arctic, with 17 glaciers regularly monitored for mass balance. In the chemical records of glaciers, a variety of substances have been determined, e.g., ions, heavy metals, or persistent organic pollutants (POPs), with the use of precise analytical techniques. However, knowledge gaps persist, preventing a formation of a reliable chemical inventory of Svalbard glaciers. Moreover, detailed studies on the deposition and transport of pollutants, rather than focusing on their presence only, are crucial future research recommendations.

Keywords: glacial catchments, anthropogenic pollutants, glacier mass balance, polar ecosystems, environmental contamination

1. Introduction

Glaciers are not only contributors to the sea level rise but can also accumulate and release pollutants [1, 2], as well as transform the chemical composition of water that originates or flows through them. Since glacial water is used by over a billion people for domestic purposes [3], both the quality and quantity of this water should be monitored. Indeed, such studies can be a vast source of knowledge on the processes in the otherwise unavailable subglacial environment. In this chapter, we concentrate on the archipelago Svalbard in the Arctic, a typical target area for xenobiotics from long range atmospheric transport (LRAT), holding an important share of the Arctic glacial ice cover. We show the ways the glaciers of Svalbard are monitored for water losses and quality changes, alongside some benefits already acquired through such studies. A new direction in the research is needed that would deepen the interpretation of the obtained monitoring data.

2. Glacier monitoring and projects implemented in the Arctic

2.1. Cryosphere

Cryosphere refers to *"the part of the Earth's crust and atmosphere subject to temperatures below 0°C for at least part of each year"* [4]. The snow, ice, and frozen ground all constitute the cryosphere, considered a source of climatic diagnosis due to its sensitivity to air temperature and precipitation changes. The most recent intergovernmental panel on climate change (IPCC) assessment [5] emphasizes also the importance of cryosphere in the Earth's ecosystem as a reservoir of solidified water. Glaciers and the great ice sheets of Greenland or Antarctica are only part of the cryosphere, as shown in **Figure 1**.

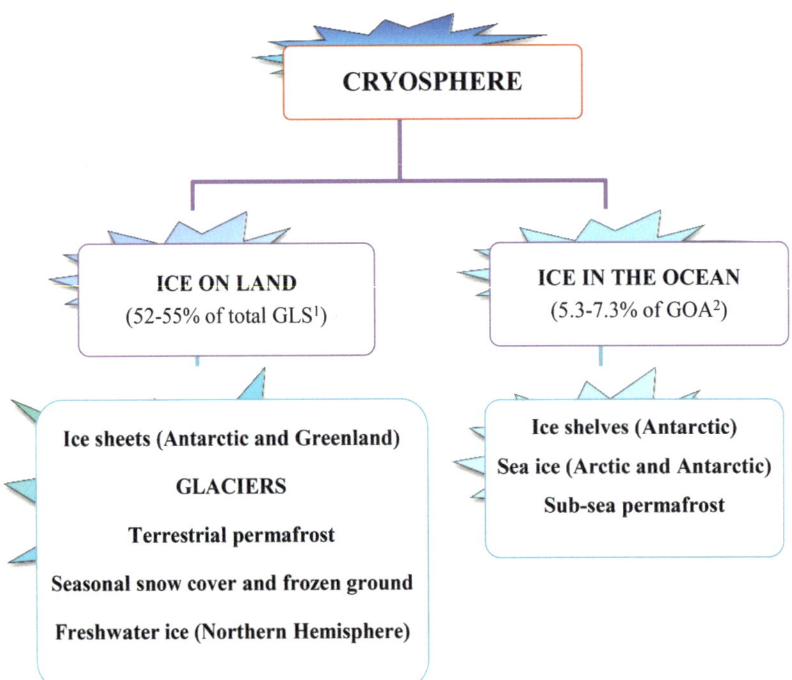

Figure 1. Division of the cryosphere and its components: [1]Global land surface: 147.6 Mkm^2, [2]Global ocean area: 362.5 Mkm^2 [5].

The global land surface covered by glaciers (0.5%) is the least abundant cryosphere component. It is referred to as "ice on land." Nevertheless, the general significance of glaciers for the sea level equivalent (0.41 m a.s.l.) is the highest among the components, except for ice sheets (Antarctic: 58.3 m, Greenland: 7.36 m). Glaciers are long-term components of the cryosphere, with a lifespan exceeding freshwater ice on rivers and lakes (seasonal) and sea ice (several years in the Arctic), but shorter than ice sheets and permafrost, surviving even millions of years [5] (**Table 1**).

Considering their contribution to the sea-level rise and the lifespan of particular components of the cryosphere, glaciers are of extreme significance for the environment as an indicator of climate change in the context of global warming.

2.2. Glacial system

Glaciers occupy 10% of the Earth's surface. As natural water reservoirs, they represent 75% of freshwater on Earth. The vast majority of the water (99.5%) is stored in the Greenland and Antarctic ice sheets. Ref. [3] has emphasized the great significance of glaciers as a source of freshwater widely used by over a billion people. Glacial waters are not only used for domestic purposes, but also for electricity production and crops irrigation (e.g., in the Alps, Himalayas). However, it is the small glaciers and ice caps of the High Arctic that have been rapidly responding to climate changes in the recent years, and therefore have contributed the most to the sea level rise [4–8].

Although the high latitude regions of the Arctic are distinguished by limited human impact and low emission from local sources, they cannot be considered free from the presence of pollutants. For example in Svalbard, the long range transport of atmospheric pollutants transmitted from Eurasian industrialized and urbanized areas may substantially affect the quality of Arctic waters, since the atmospheric deposition is one of the main factors (next to rock-water interaction) controlling water chemistry in this polar region [9–13].

Due to glacial drainage and the processes by which glaciers are formed, they are an important element in the global water cycle. The accumulation of water as snow and its gradual release in the liquid form determine the importance of glacial controls upon the drainage characteristics of partly glaciated catchments [6].

Element of the cryosphere	Lifespan
Snow	A day to several months
River and lake ice	Several days to several months
Sea ice	Several days to almost a year
Glaciers and ice caps	Months to a century
Frozen grounds	A day to a millennium
Ice sheets, ice shelves	Days to a millennium

Table 1. Lifespan of selected elements of the cryosphere.

Glaciers develop when snow accumulates over a period of several years, and then gradually transforms into firn (at least 1 year old snow) to finally turn into ice. The ice flows downward due to the force of gravity. Snow accumulation predominantly depends on the climate conditions and topographic characteristics [4, 5, 14, 15]. When the accumulation process (snowfall) prevails over the ablation processes (iceberg calving, surface melting, and runoff, melting under floating ice shelves), glaciers gain mass [4, 5, 14, 15].

An important typical feature of glaciers is the circulation of mass, i.e., ice, snow, water, and mineral matter, as well as the circulation (exchange) of energy manifested in accumulation, the glacier movement, and ablation. The processes are determined by external factors. They also substantially affect the environment, making the glacier a dynamic, open system [16]. The glacier system is fed by and releases various forms of energy and mass, which are subject to further movement and/or transformation inside it. **Figure 2** shows the relations between the entry and exit elements.

The mass movement is determined by the force of gravitation. Energy transformations and movement are accompanied by complex processes within the glacier. As a result of differences in the mass balance, the uniform glacial system is divided into two spatial subsystems, namely the accumulation and ablation system, separated by the equilibrium line [16, 17]. Maintaining balance within the glacial system is only possible when the balance elements (entry and exit elements of the system) are equal, and the mass flow through the equilibrium line is even. Any disturbance in the balance causes a response of the system in the form of feedback loops. An example of such a process is an increase in accumulation, which causes an increased flow of ice

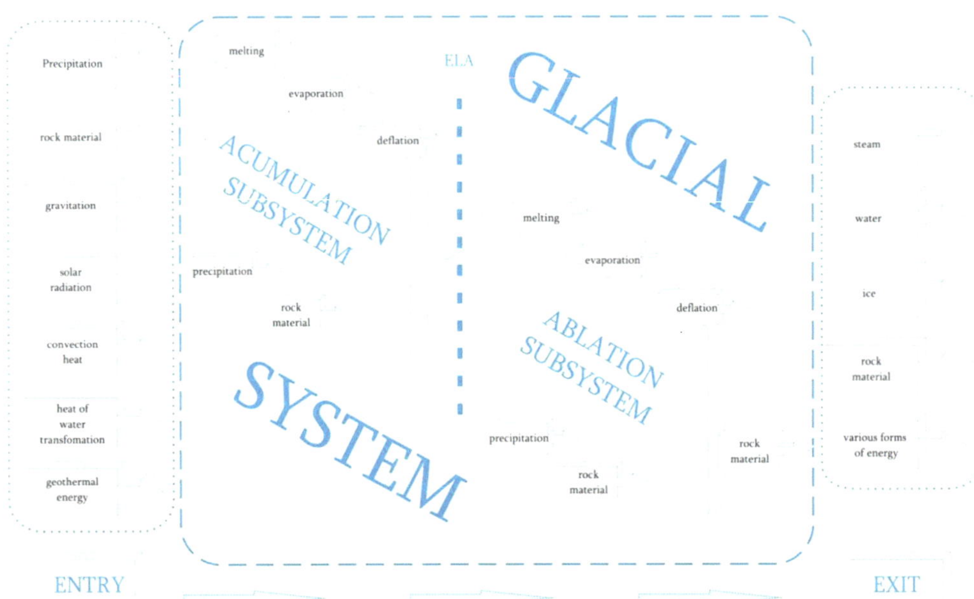

Figure 2. Schematic model of the glacial system.

mass through the equilibrium line, contributing to the advance of the glacier terminus, an increase in ablation, and reduction in the lower part of the glacial system.

The dynamic open glacial systems substantially affect climate at the global scale. They are also excellent indicators of climate fluctuations. The response of glaciers to climatic changes varies depending on their morphological features and internal thermal structure [14, 16].

Polythermal glaciers, i.e., glaciers with a complex thermal structure, are particularly good indicators due to their response to changing climate characteristics, as they are developed not only as a result of varied air temperature, but also by variable amount and structure of precipitation. In contrast to glaciers with cold thermal regime, the internal hydrothermal structure of polythermal glaciers is determined not only by solid, but also by liquid precipitation [18].

2.3. The beginning of world glacier monitoring

Glacier monitoring has a history dating back to the nineteenth century (**Figure 3**). The father of glacier monitoring was François-Alphonse Forel, the first scientist to observe changes in Alpine glaciers. The first international initiative emerged during the sixth International

Figure 3. Most important dates in the early history of glacier monitoring.

Geological Congress in Zurich. Since then, scientists have been collecting information on changes in selected glaciers and performing detailed surveys of their tongues on a regular basis. The data were enriched by the indigenous knowledge on earlier glacier stages, provided by the mountain people. At the early stages of the research, it focused on glacier fluctuations, therefore only data on front variations were published. Since 1940, information regarding mass balance has been included in publications. The need for a worldwide inventory of the existing ice and snow masses was recognized just after the declaration of the International Hydrological Decade (1965–1974) by the United Nations Educational, Scientific, and Cultural Organization (UNESCO). This resulted in the establishment, under the auspices of UNESCO, of the first international network called the Permanent Service on the Fluctuations of Glaciers (PSFG). Worldwide glacier monitoring has been rapidly evolving since then, and in 1975, the Temporal Technical Secretariat for the World Glacier Inventory (TTS/WGI) was established. Its role was to collect and periodically publish glacier inventory and fluctuation data. The tasks of TTS/WGI and PSFG were taken over by the World Glacier Monitoring Service (WGMS), established in 1986. The first status report of glaciers inventory, published in 1989, includes information on their geographic location, area, length, orientation, elevation, and classification of morphological type and moraines. The data were mainly based on aerial photographs, maps, and satellite images. Since 1995, when project Global Land Ice Measurements from Space (GLIMS) was launched, the data have also been collected from optical satellite instruments such as the Advanced Spaceborne Thermal Emission and reflection Radiometer (ASTER) [19–21]. The collaboration of WGMS with the US National Snow and Ice Data Center, initiated in 1998, resulted in the first data inventory available online via the website of the National Snow and Ice Data Center (NSIDC) already a year later [20, 22, 23]. The most important dates and events in the early history of worldwide glacier monitoring are provided in **Figure 3**.

2.4. The organization of the glacier monitoring system

The establishment of the Global Terrestrial Observing System (GTOS) in 1996 was a consequence of the Second World Climate Conference held in 1990. The conference called for the establishment of a coordinated monitoring system (**Figure 5**). The Terrestrial Observation Panel for Climate (TOPC) was established within GTOS. The global observing strategy was subsequently designed. It permits introducing all variables essential for the climate (e.g., river discharge, groundwater, lakes, glaciers, and ice caps) related to monitoring systems to the Global Terrestrial Network (GTN). As a result, the Global Terrestrial Network for Glaciers (GTN-G) was established in 1998. GTN-G is responsible for collecting standardized data on the current state of glaciers. Since its establishment, it has been run by WGMS with the assistance of NSIDC and GLIMS. The monitoring system is under the supervision of several worldwide organizations presented in **Figure 4** [22, 24].

2.5. Glacier research projects in Svalbard

Due to the strong response of glaciers to climate change, their great importance for sea-level rise, and impact on the environment, many international research programs and projects have been conducted in the Arctic, including Svalbard. Research projects, unlike monitoring, include innovative testing of new methods and techniques and have a typical duration from

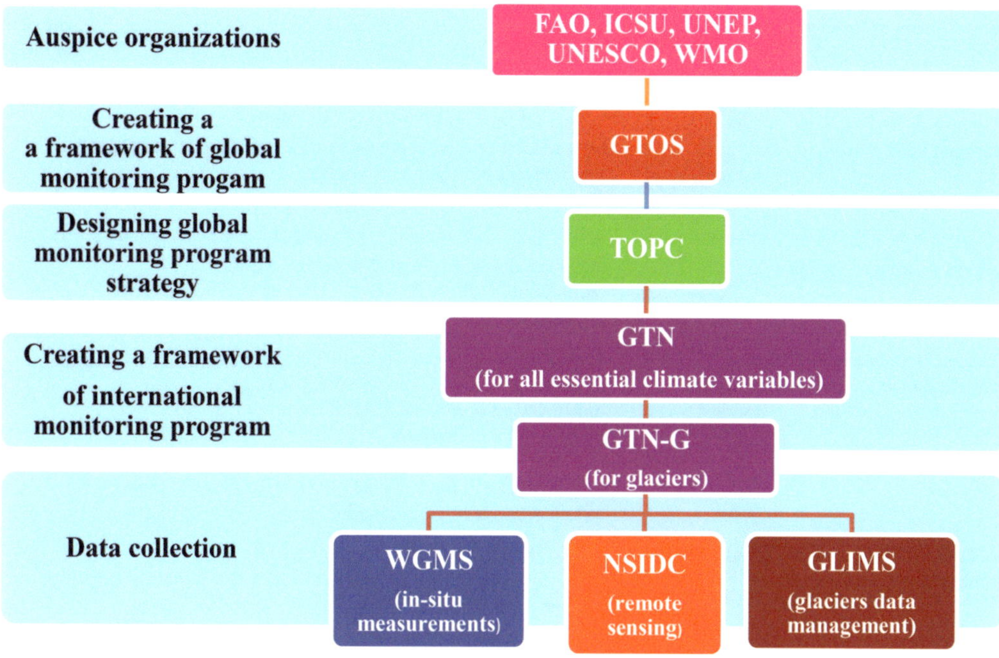

Figure 4. Major international organizations and their role in glacier monitoring.

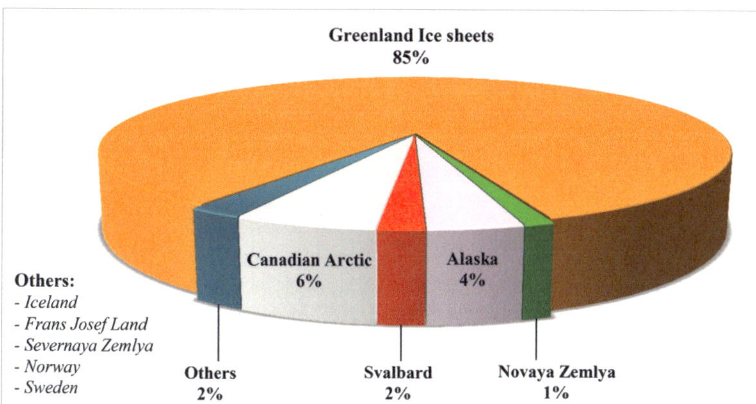

Figure 5. Percent contribution of 10 Arctic regions covered by extensive glaciation [4].

3 to 6 years, and are funded from different sources. Examples of such research projects regarding Svalbard glaciers are listed in **Table 2** [25–28]. A great number of projects is interdisciplinary, concerning both glaciology and glacial hydrology. Some are also related to meteorology (e.g., CRYOMET) and seismology (e.g., SEISMOGLAC). The vast majority of research projects is associated with the response of the cryosphere to global warming and climate change.

Project (years of implementation)	Scope of research	Source of funding
"GLACIODYN" The dynamic response of Arctic glaciers to global warming (2007–2010)	1. Current mass budget of each glacier (including calving); 2. Subglacial processes (hydrology and sliding interactions); 3. New models of calving processes (numerical models including functions of sliding and calving); 4. Prediction of glacier response to climate change scenarios.	RCN
"SvalGlac" Sensitivity of Svalbard glaciers to climate change	1. Measurements of mass budget, glacier flow velocity, glacier thickness and hydrothermal structure, weather; 2. Studies on actual glacier topography and shallow ice cores (for the past climate reconstruction).	ESF
"Ice2sea" Estimating the future contribution of continental ice to the sea-level rise (2009–2013)	1. Studies of key processes in longer-lived elements of the cryosphere (mountain glacier systems, ice caps, ice sheets); 2. Improvement of satellite determinations of current changes in continental ice masses; 3. Development of a detailed forecast of the contribution of continental ice to the sea-level rise over the next 200 years by means of ice-sheet/glacier models.	ERC
"ICEMASS"ERC advanced grant global glacier mass continuity	1. Data collection and analysis regarding glacier thickness changes, and converting the data to a global glacier mass budget; 2. Estimation of the current sea-level contribution from glaciers; 3. Studies on glacier mass changes reflecting climate change patterns; 4. Examination of the impact of glacier imbalance on river run-off.	ERC
"CRYOMET" Bridging models for the terrestrial cryosphere and the atmosphere (2012–2015)	1. Validation of the polar weather research and forecasting model (WRF) land surface scheme; 2. Collection of cryosphere data sets constituting variables with Polar WRF; 3. Studies on further probabilistic downscaling of snow cover by means of snow distribution models; 4. Tests of upscaling schemes for the surface energy balance in polar WRF (cryosphere-atmosphere feedbacks).	RCN
"SEISMOGLAC"-Seismic monitoring of glacier activity on Svalbard (2012–2015)	1. Studies on the relation between glacial process and seismicity; 2. Finding the source location of ice quakes; 3. Use of automatic pattern recognition methods to classify their signals.	RCN

Abbreviations: RCN: Research Council of Norway, ESF: European Science Foundation, ERC: European Research Council.

Table 2. International research projects in Svalbard within the framework of which glaciers are studied.

3. Svalbard glaciers and climate warming

Part of the cryosphere of the northern hemisphere categorized as "ice on land" is distributed irregularly in the Arctic. Therefore, glaciers and ice caps are subject to different climatic conditions. In Ref. [5], 10 regions of the Arctic are specified as covered by extensive glaciation. Together they occupy an area of 1,972,600 km^2. The percent contribution of each of them is shown in **Figure 5**.

Svalbard archipelago is the most glaciated region of the European Arctic. The area of its glaciation (approximately 36.6 km^2) is substantially higher than that of Norway and Sweden (approximately 3.1 km^2), Iceland (approximately 10.9 km^2), Franz Josef Land (approximately 13.7 km^2), and Novaya Zemlya (approximately 23.6 km^2).

The response of the Greenland ice sheet to climate change is slower, because more than a half of its surface experiences temperatures well below the freezing point during the entire year. Changes in temperature or precipitation cause a more rapid response in smaller glaciers and ice caps, which are more sensitive [4]. Throughout the Arctic, except for Russian Arctic, the mass balance (difference between annual mass gain and annual mass loss) is only monitored on 27 glaciers. Four of them are located on Svalbard (Midre Lovenbreen, Austre Broggerbreen, Kongsvegen, and Hansbreen) [29].

The Svalbard archipelago includes four main islands (Spitsbergen, Nordautslandet, Edgeøya, and Barentsøya) and occupies an area of 62,248 km^2. Approximately, 60% of the Svalbard archipelago is covered with ice. The glacier inventory of Svalbard amounts to 1615 glaciers and ice caps, of which 17 are under permanent or periodic mass balance research.

Sixty percent of Svalbard glaciers are terminating in the sea at calving ice-cliffs. Ref. [7] has emphasized that due to calving, the annual specific mass loss of Svalbard glaciers is much higher than from the Greenland ice sheet and seems to be the highest in the Arctic. Each of the main islands of the archipelago represents a different type of landscape (more detailed information is provided in **Table 3**). On Spitsbergen, the largest island of the archipelago, 90% of glaciers are considered polythermal (subpolar) [8, 30, 31].

Although the dominant component of Spitsbergen landscapes is rugged mountains with glaciers, its eastern part is covered with several large ice caps. Together with ice caps from three other islands, also calving into the sea, they all develop a calving ice front with a total length of approximately 1000 km. The total volume of the ice masses of Svalbard is estimated at 7000 km^3 [7, 32].

3.1. Role of glaciers in the Svalbard environment

Glaciers occur in places where climatic and topographic conditions favour snow accumulation. Their role may be considered both at the global scale and at the regional scale, as shown in **Figure 6**.

Islands	Landscape	Area of glaciation
Spitsbergen	Steep, rugged mountains	~22,000 km^2 of glaciers ~14,600 km^2 of ice caps
Nordautslandet	Two largest single ice bodies within Svalbard	~2450 km^2 (Vestfonna ice cap) ~8000 km^2 (Austfonna ice cap)
Edgeøya and Barentsøya	Plateau-type terrain	~2800 km^2 of low altitude ice caps

Table 3. Dominant types of landscapes and extent of glaciation of the main Svalbard archipelago islands [8, 32].

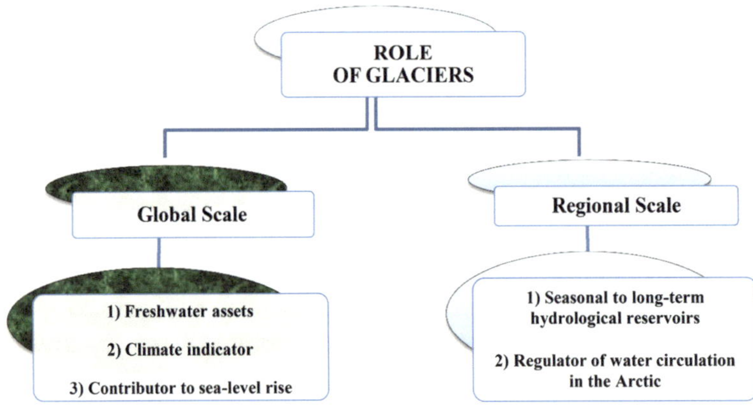

Figure 6. The role of glaciers in the environment [5, 32].

Glaciers adjust their size in response to changes in climate, e.g. in temperature and precipitation. Therefore, they are considered very sensitive climate indicators. Changes in their size or shape may be observed over several decades or even several years. Svalbard glaciers also have a contribution in the sea level rise, estimated at 4% of the total contribution of smaller glaciers and ice caps. The contribution of the archipelago corresponds to the ratio between the glaciated area of Svalbard and the global surface covered by glaciers and ice caps [4, 8].

For the Arctic environment, with a fragile homeostasis, the regional role of glaciers is significant [10, 32]. Glaciers are the most visible component of the Svalbard environment. Due to this, they can also be considered a major geomorphological factor of the entire archipelago [32]. They respond the fastest and strongest to climate changes among all environmental components, and are a major regulator of water circulation in the Arctic [33]. Ref. [32] has emphasised the role of glacier runoff in Svalbard, a factor affecting not only the hydrology of rivers, but also circulation in the neighbouring seas and fjords, due to changes in stratification within the water column. Local climate and biota may also be influenced by changes in the glacial runoff, affecting the sea ice conditions of the archipelago. Even deep-water production close to the shelf of Svalbard may be influenced by a rapid discharge of freshwater from glaciers [32].

4. Glacier water chemistry and the origin of chemical additions

The chemical composition of glacial meltwaters in Svalbard has been subject to increased interest in recent years [34–37]. Waters originating from glaciers affect the quantity and quality of water delivered to the environment in glaciated catchments, which also plays an important role in ice mass dynamics [13]. Furthermore, glaciated catchments regulate the biogeochemical circulation of nutrients, and influence the cryosphere-atmosphere interactions [38].

Although the high latitude regions of the Arctic experience very limited human impact and low emission of local origin, they cannot be considered free from pollutants any more. The long-range atmospheric transport from the regions of Eurasia with higher emissions may influence the Arctic water quality, making atmospheric deposition one of the main factors (alongside rock-water interactions) controlling water chemistry of the polar regions. Due to high rates of chemical weathering and minimal human impact in glaciated areas, they constitute an environment almost ideal for studying water-rock interactions [11–13]. Hydrochemical data on proglacial waters provide explanation of water drainage pathways through glaciers and estimations of chemical weathering rates [34, 39].

Ref. [12] has emphasized the specific conditions of the Arctic environment, such as: *"(1) relatively short water-rock contact time, (2) cold temperatures, (3) thin soils, and (4) lack of vegetation,"* which reduce the activity of geochemical processes, including chemical weathering. However, the contact of water with eroded glacial debris and the abundance of soluble rocks such as carbonates and sulphides tend to considerably enhance such activity. In addition to the chemical weathering of rocks and the atmospheric deposition, other factors potentially influence dissolved solute concentrations, these are *"(1) discharge conditions at the time of sampling, (2) inputs and outputs from the soil exchange pool, (3) uptake of organic nutrients by biomass, and local variations in non-living organic material (humus), and (4) changes in the topography and soil development"* [12, 13].

The ionic composition of glacial meltwater varies due to different types of its transit through the glacial system, and the duration of chemical weathering reactions supplying solutes to such waters. The variety of glacial processes, and consequently chemical weathering, is strongly influenced by the thermal regime of glaciers. Meltwater in contact with the bedrock is present in temperate and subpolar glaciers. The acquisition of solute derived from chemical weathering occurs at the glacier bed during the transit of meltwater through two types of drainage systems: distributed and channelised. The distributed drainage involves linked cavities or porous flow through permeable subglacial sediments, and is mainly fed by snowmelt or slow transit of meltwater. This system of drainage is characterised by high water pressure and long residence time. The rock-water contact area is high. The channelised drainage system is fed by ice melt, mixed with waters from the distributed system to produce bulk meltwater. This system rapidly drains high volumes of water from beneath the glacier. The chemical reactions occurring on the water-rock interface in the glacial system depend on the type of drainage and their changeability during the ablation season [13, 40].

The most important mechanism of chemical rock weathering is acid hydrolysis. Ref. [13] has emphasized that dissolved anion signature of the meltwater indicates the source of protons necessary to drive acid hydrolysis reactions. Furthermore, Ref. [13] has listed the sources of protons such as (1) dissociation of atmospheric CO_2 [Eq. (1)], (2) sulphide oxidation [Eq. (2)], and (3) oxidation of pyrite [Eq. (3)]. The latter is often coupled with carbonate dissolution.

$$CaCO_3(s) + H_2CO_3(aq) \leftrightarrow Ca^{2+}(aq) + 2HCO_3^{-}(aq) \tag{1}$$

$$4FeS_2(s) + 14H_2O(l) + 15O_2(aq) \leftrightarrow 16H^{+}(aq) + 8SO_4^{2-}(aq) + 4Fe(OH)_3(s) \tag{2}$$

$$\begin{aligned} &4FeS_2(s) + 16CaCO_3(s) + 14H_2O(l) + 15O_2(aq) \leftrightarrow 4Fe(OH)_3(s) + 8SO_4^{2-}(aq) \\ &+16Ca^{2+}(aq) + 16HCO_3^{-}(aq) \end{aligned} \tag{3}$$

The relative proportions of HCO_3^- and SO_4^{2-} in the bulk outflow reflect the dominance of the major sources of aqueous protons driving subglacial weathering reactions. Ref. [13] has assumed that, when using the C-ratio $[HCO_3^-/(HCO_3^-+SO_4^{2-})]$, a value of 1 signifies weathering by carbonation reactions, while a value of 0.5 reflects coupled sulphide oxidation and carbonate dissolution.

4.1. Pollutants examined in the catchments of Svalbard glaciers

Next to the natural chemicals from rock-water contact, human activity also contributes chemicals to Arctic waters, despite the distance of thousands of kilometres between the Arctic and the industrial and agricultural areas. During the last two decades, pollutants continued arriving into the Arctic, and despite their decreasing or steady atmospheric levels [41], their negative impact on the polar environment remains an important concern [42–49].

The Svalbard archipelago is different from the other Arctic regions. Due to its geographical location and specific climate conditions, it is particularly exposed to the accumulation of a wide range of chemical substances recognised as pollutants [9, 10]. Its relatively short distance from continental Europe, the location of the archipelago in the gap between the continents surrounding the Arctic Basin, and its landscape dominated by rugged mountains with glaciers, make it conducive to the accumulation of pollutants on its glaciers. Moreover, ocean and wind currents contribute to the transport of pollutants from lower geographic latitudes. In combination with low temperatures, this results in Svalbard and its glaciers becoming a sink for xenobiotics [50–54]. Although the levels of multiple pollutants such as heavy metals and many POPs contained in various elements of the living and inanimate environment are well known [10], knowledge on the fate of pollutants in Svalbard glaciers is still scarce.

Many scientific studies discuss the issue of the contamination of the Arctic environment. A vast number of publications concern the content of xenobiotics detected both in the living organisms (e.g. [46, 55–59]) and in the inanimate environment [60–64]. In Ref. [10], levels of pollutants present in samples collected in the Svalbard archipelago are discussed in detail. This paper focuses on the literature directly related to the presence of a wide range of chemicals recognized as pollutants in glacial catchments. The majority of research on the chemistry of glacier catchments is performed on Spitsbergen, the largest island of the archipelago (**Figure 7**).

The research site locations are directly related to the occurrence of the warm West Spitsbergen Current, considerably affecting the climate of the western coast of Spitsbergen. The warm waters limit the sea-ice development, which makes this area easier available for research activities. This is evident in the contribution of individual fjords, with the only representant of the eastern side of the island being Woodfjorden. Moreover, due to the cold East Spitsbergen Current, the east coast is dominated by several large ice caps [7, 10, 32].

Three main types of glacial catchments on Svalbard may be distinguished. The first two types involve the glacier terminus ending in the sea. In the first case, the glacier basin covers the coastal valley, and in the second case, the basin reaches into the centre of the island, covering large glaciated valleys. The third type of a glacial catchment is distinguished by the glacier terminus ending on land [78]. The glacier moraine is located in front of the glacier terminus, at

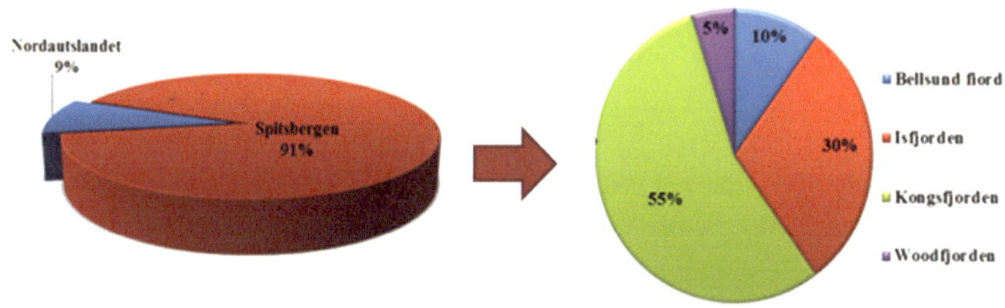

Figure 7. Places of conducting chemical research in glacial catchments in Svalbard, including the contribution of particular fjords of the Spitsbergen island [12, 34, 35, 38, 40, 49, 55, 65–77].

a certain distance from the seashore. Ablation water leaving the glacier flows through the glacier moraine and into the fjord via a number of channels developing a river system between the glacier and the fjord. Various types of surface water samples can be collected and examined depending on the type of catchment. According to the literature, glaciers representing the latter type of glacial catchments are subject to most frequent research activities (**Figure 8**). The evaluation was based on selected scientific articles, cited in **Tables 4**, **5** and **6**.

A vast majority of publications [12, 34, 40, 49, 75–77] focused on water from glaciers (proglacial, supraglacial, subglacial, and cryoconite waters). However, some include also direct

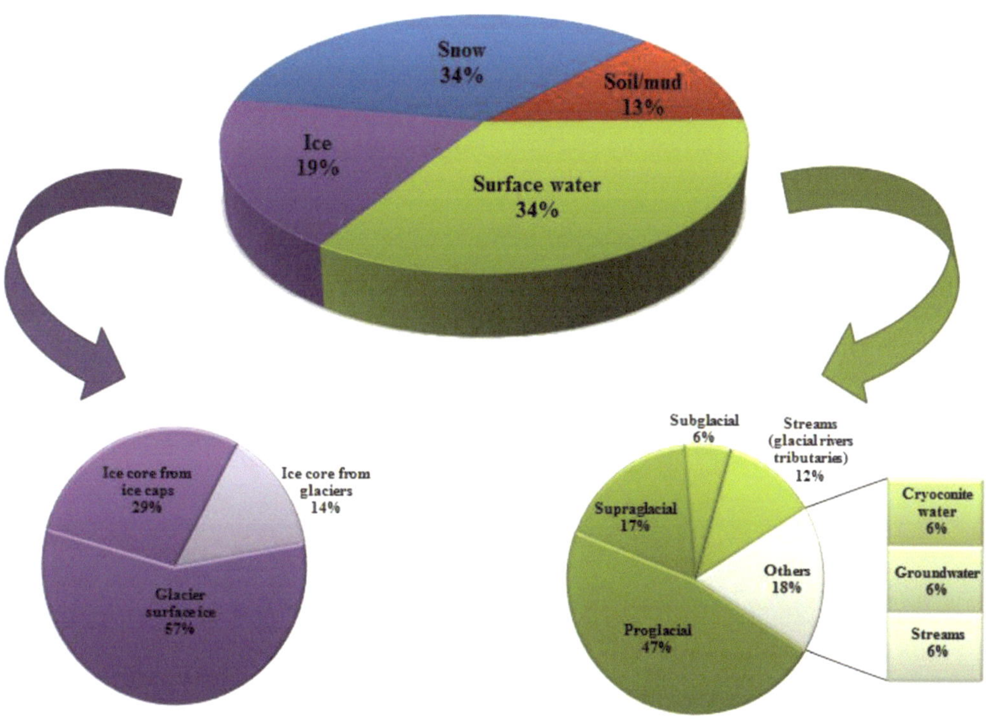

Figure 8. Contribution of different type of samples examined on the Svalbard archipelago [12, 34, 35, 38, 40, 49, 55, 65–77].

tributaries of glacial rivers [12, 76], as well as other streams and groundwaters functioning in glacier basins [12, 38]. A smaller number of studies involves the analysis of ice samples. In prevalence, the examined ice was collected from the surface of glaciers rather than from drilled ice cores, and this sampling strategy may be driven by the predominance of polythermal regime among Spitsbergen glaciers (90%). The percolation of water and chemical substances in this thermal regime disturbs the original depositional sequence of chemical composition, making it difficult to analyze their accumulation in glaciers over time. Therefore, the examined Svalbard ice cores originate usually from ice caps. Only in Ref. [49], authors analyze pollutants in ice cores collected from the polythermal glacier of Longyearbreen. Snow samples for analysis are collected from the surface of glaciers and their surroundings in nearly equal proportion. Substantially, more sediment samples from cryoconite holes [72, 77, 79] on glaciers are subject to research than soil samples collected in glacier catchments.

Projects listed in **Table 2** mainly focus on glaciological investigations [80–84]. Some of them are associated with the impact of climate change on cryosphere components and the modelling of possible cryosphere-climate interactions [8, 85]. Many scientific works also focus on the presence of pollutants such as heavy metals or POPs in biotic samples [50, 55, 61, 86, 87]. The majority of the research is related to biochemistry, and refers to the processes of bioaccumulation and biomagnification of pollutants within the marine or terrestrial food webs. Publications concerning abiotic samples collected from glacier catchments of Svalbard focus on types of research presented in **Figure 9**. The evaluation was based on selected scientific articles cited in **Tables 4**, **5** and **6**.

According to the literature review, the majority of conducted research concerns the fate and transport of pollutants in the abiotic environment. These publications mostly refer to levels of selected metals (e.g., Al, Hg) or POPs (e.g., PCN, PCBs, PFOA, PFOS, DDD, DDE, DDT) in snow [49, 67, 69–71] and ice samples [49, 55, 73, 74]. Ref. [49] discusses the effect of pollutants

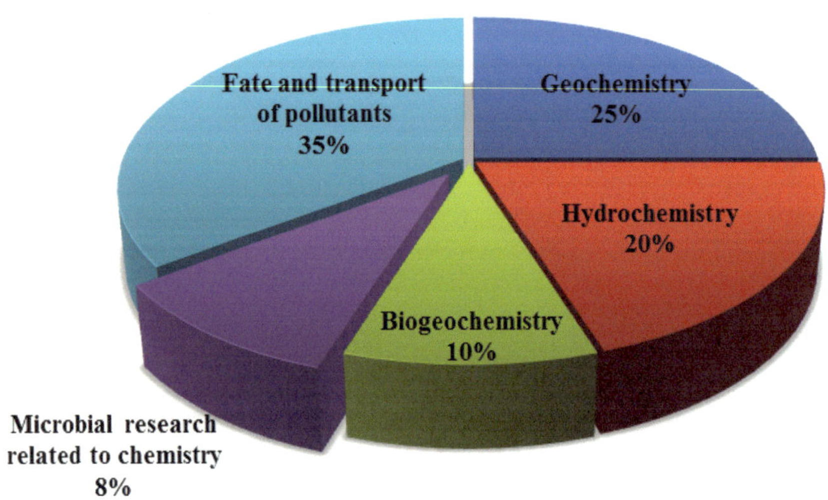

Figure 9. Types of research performed on inanimate samples collected in the Svalbard archipelago [12, 34, 35, 38, 40, 49, 55, 65–77].

present in snow, ice, and surface water samples (i.e., supraglacial lake or river and sea water) collected throughout the glacial catchment, starting from the top of the glacier and ending in the fjord waters. This is the only work providing an insight into the transport of anthropogenic pollutants through almost all of the elements of the glacial catchment. A considerable number of publications focus on the chemical weathering process [12, 34, 35, 38, 75]. Others concern seasonal changes in hydrochemistry [68, 76], or compare the hydrochemistry of inanimate samples collected in different parts of the environment [66, 72]. Such works mainly present results of analysis of inorganic ions (e.g., K^+, Na^{2+}, Mg^{2+}, F^-, Cl^-, NO_3^-, NO_2^-). A smaller number of studies concerns biogeochemistry [40, 65] or microbiology related to the fixation of nitrogen on glaciers or carbon cycle [77, 79].

Since data from long-term chemical monitoring of glaciers are scarce and rarely published in full, we collected here an inventory of shorter published measurement series or important datasets that can be treated as a proxy of the current state of the glacial chemical monitoring in Svalbard. First, we present an overview of the techniques and equipment used for the determination of a wide range of analytes studied in the environmental samples from glacial catchments of the Svalbard Archipelago. We have divided the data into three categories: snow (**Table 4a**), ice (**Table 4b**) and surface water (**Table 4c**). According to the literature review, ion chromatography (IC) is the analytical method that is used most frequently for the determination of not only inorganic ions, but also other pollutants (e.g., methyl-sulfonic acid and glutaric acid). The determination of the concentration of metals in the environment usually involves the methods of flow injection analysis (FIA) or atomic absorption spectroscopy (AAS). The determination of organic pollutants, which are highly detrimental for Arctic biota, is performed by means of gas chromatography (GC), usually coupled with mass spectrometry (MS) in different resolution modes (low resolution, high resolution). Inorganic ions and metals are the most frequently determined analytes in almost all of the elements of glacial catchments (snow, ice, water, soil, and cryoconite). Research involving the determination of persistent organic pollutants (e.g., DDD, DDT, PCBs, PCNs, HCH, HCB) is conducted very rarely in the glacier catchments. These dangerous chemical compounds are usually determined in snow and ice samples (ice cores and surface ice) collected in the glacial catchment, where they reflect contribution of long-range atmospheric transport and their history of accumulation.

Except water samples, other abiotic material has also been investigated in the glacial catchments of Svalbard, especially rock material of different types. For example, cryoconite sediment has been analysed for its nutrient content (for DIN, TIN and TN, using Bran and Luebbe Autoanalyzer 3, [77]) or heavy metal concentration (Fe, Mn, Zn, Pb, Cu, Cd; by voltamperometric and spectrophotometric method, [72]). Similar parameters to water samples were established in soils, especially pH [34, 65], inorganic anions (Cl^-, NO_3^-, SO_4^{2-}, HCO_3^-) and cations (Na^+, K^+, Mg^{2+}, Ca^{2+}) [34, 65], SiO_2 concentration [34], and organic carbon and nitrogen [65]. The methods used in the mentioned studies matched those used for snow, with the exception of the Fisons NCS analyser application for organic carbon and nitrogen.

In **Tables 5a**, **5b** and **5c**, we present the published chemical concentration data from the samples described in **Tables 4a**, **4b** and **4c**, respectively. Most studies concerned watercourses, and there the highest variability of chemical parameters was found. Ice samples have shown

Determined compound(s)/parameters	Analytical method/apparatus		References
pH			[65]
	pH meter		[66]
	Heito pH meter (Paris)		[67]
	Orion SA 250 portable meter with Ross combination electrode		[34]
	Orion 290a portable pH meter with Ross combination electrode		[68]
EC	conductivity meter		[66]
Anions (Cl^-, Br^-, NO_3^-, SO_4^{2-})	IC	Dionex ion chromatography	[65]
		DionexR 2100	[66]
		Dionex DX100	[40]
		IC, colorimetric method	[38]
		Dionex ICS 3000	[67]
		Dionex 4000i	[34, 68]
		Dionex ICS-1100	[49]
HCO_3^-	Titration (0.01 M HCl)		[68]
	Titration (1 mmol HCl)		[34]
	Titration (10^{-3} mol/L H_2SO_4)		[65]
Cations (Na^+, NH_4^+, K^+, Mg^{2+}, Ca^{2+})	AAS		[34, 38, 65]
	ICP-OES		[66]
	FIA		[40]
	IC	Dionex ICS-3000	[67]
		Dionex 4000i	[68]
		ICS-1100 Dionex	[49]
Metals	Al_{total}	AAS	[65]
	Hg_{total}	CVAFS	[67]
		ICP-SFMS	[69]
	$Hg_{reactive}$	ICP-QMS	[69]
	MMHg (monomethylmercury)	AFS	[67]
S	ICP-OES		[66]
Si-Si$(OH)_4$	FIA		[65]
SiO_2			[34]
Si			[68]
MSA(methyl-sulfonic acid), Glut (glutaric acid)	IC (Dionex ICS 3000)		[67]
	GC-MS-EI-SIM		[70]

Determined compound(s)/parameters	Analytical method/apparatus	References
∑PCB9, α-HCH, γ-HCH, ∑DDT, HCB, chlordane (cis- or trans-)		
∑PCN, ∑PCB	HRGC-LRMS	[71]
PFOA, PFOS	LC-MS/MS	[49]

Table 4a. Literature data on the analytical techniques and equipment used for the determination of a wide range of compounds in the snow samples (snowfall, surface snow, snowpack) collected in the glacial catchments of the Svalbard archipelago.

Determined compound(s)/parameters	Analytical method/apparatus		References
pH/EC	pH/conductometer CPC-411 by Elmetron		[88]
Anions (Cl^-, NO_3^-, SO_4^{2-})	IC, colorimetric method		[38]
	IC (ICS-1100 Dionex) IC (ICS-3000Dionex)		[49, 88]
Cations (Na^+, NH_4^+, K^+, Mg^{2+}, Ca^{2+})	AAS		[38]
	IC (ICS-1100 Dionex) IC (ICS-3000Dionex)		[49, 88]
Metals		AAS	[55]
	Zn, Mn, Cu, Fe, Ni, Cr, Pb, Cd, Co	Voltamperometric and spectrophotometric method	[72]
	Li, Be, B, Al, V, Cr, Mn, Fe, Co, Ni, Cu, Zn, As, Se, Rb, Sr, Cd, Cs, Ba, La, Ir, Pb, Th, U	ICP-MS	[88]
PFOA, PFOS	LC-MS/MS		[49]
Aldrin, Dieldrin, α-HCH, Heptachlor, Heptachlor epoxide, Methoxychlor, Chlorpyrifos, Dacthal, Methyl-parathion	GC-LRMS-EI GC-ECD		[73] [74]
γ-HCH, α Endosulfan, β Endosulfan, Diazinon, Dimethoate, Disulfoton, Imidan, Terbufos, Alachlor, Pendamethalin, Desethyl atrazine	GC-LRMS-EI GC-LRMS-ECNI		[73] [74]
Endrin-aldehyde	GC-LRMS-EI		[73]
Endrin, Endrin-ketone, Cis-nonachlor, Trans-nonachlor, o.p'-DDD, p.p'-DDT (L), p.p'-DDE, γ-chlordane, α-chlordane,	GC-ECD		[74]
Endosulfan sulphate, Metolachlor, Trifluralin, Metribuzin	GC-LRMS-ECNI		[74]
DOC	TOC analyser (Shimadzu)		[88]

For acronyms see list in the beginning of the article.

Table 4b. Literature data on the analytical techniques and equipment used for the determination of a wide range of compounds in the ice samples (glacier surface, ice cores from glaciers and ice caps) collected on the glaciers of Svalbard.

Determined compound(s)/parameters	Analytical method/apparatus		References
pH			[37, 38, 65]
	Orion SA 250 portable meter with a Ross combination electrode		[34]
	Jenco pH-meter		[35]
	Orion (Thermo Scientific), WPA (Cambridge, UK) or VWR pH meter		[36]
EC	CC-317 conductivity meter		[35]
Anions (Cl^-, NO_3^-, PO_4^{3-}, SO_4^{2-})	IC	Dionex 4000i	[75]
		Dionex Ion Chromatography	[65]
		Dionex DX100	[40, 36]
		IC, colorimetric method	[38]
		Dionex 4000i	[34]
		Metrohm Compact IC 761	[35, 37]
		Dionex DX-120	[12]
		Dionex ICS-90	[76, 36]
HCO_3^-	IC	Dionex 4000i	[75]
		Dionex DX-120	[12]
	Titration (10^{-3} molc/L H_2SO_4)		[65]
	Colorimetric titration		[38]
	Titration (1 mmol HCl)		[34]
	Titration (10 mmol HCl)		[36]
	Titration (0.02 M HCl)		[35]
	Titration (Metrohm 702 SM Titrino)		[37]
N-NO_2^-	IC	Dionex DX-120	[12]
N-NO_3^-		Dionex Ion Chromatography	[65]
		Dionex DX-120	[12]
N-NH_4^+		Dionex DX-120	[12]
	FIA		[65]
Cations (Na^+, NH_4^+, K^+, Mg^{2+}, Ca^{2+})	IC	Dionex 4000i	[75]
		Dionex ICS-90	[76, 36]
		Metrohm Compact IC 761	[37]
	AAS		[12, 34–36, 38]
	FIA		[40]
Metals (Al_{total}, Fe, Mn, Zn, Pb, Cu, Cd)	AAS		[65]
	FIA		[34]
	Voltamperometric method, spectrophotometric method		[72]
Si	Colorimetric method (Skalar Autoanalyser)		[76]

Determined compound(s)/parameters	Analytical method/apparatus	References
Si-Si(OH)$_4$	FIA	[65]
SiO$_2$		[34]
	Spectrometry using the method of reduction to molybdenum-blue	[35, 36]
	Perkin Elmer ICP-OES Plasm 40 spectrometer	[37]
DOC	LABTOC Carbon Autoanalyser	[65]
DOC, DON	Shimadzu Total Organic Carbon (TOC)/Total organic nitrogen (TON)-V analyzer	[38]
DIN	Bran and Luebbe Autoanalyzer	[77]
DIC	Estimated from charge balances	[76]
PFOA, PFOS	LC-MS/MS	[49]

For acronyms see list in the beginning of the article.

Table 4c. Literature review of the analytical techniques and equipment used for the determination of chemical parameters in the surface water samples (glacial waters, streams, springs, cryoconite water) from glacial catchments of Svalbard.

Determined compound(s)/parameters		Identified level/range			References
pH		4–6.82			[34, 65–68]
EC		[µS cm^{-3}] 6.1–80.4			[66]
Anions		[µmol L^{-1}]	[mg L^{-1}]	[µEq L^{-1}]	
	Cl^-	0.9–553	0.2–20.7	<LOD-2400	[34, 38, 40, 49, 65–68]
	Br^-	<LOD-0.90	–	–	[67]
	NO_3^-	0.1–3.9	0.01–0.162	<LOD-7	[34, 38, 40, 49, 66–68]
	SO_4^{2-}	0.4–34.5	<LOD-2.82	<LOD-240	[34, 38, 40, 49, 66–68]
	HCO_3^-	[µmol L^{-1}]		[µEq L^{-1}]	
		57.1–195		11–900	[34, 65, 68]
	N-NO_3^-	[mg L^{-1}]			
		0.01–0.02			[65]
Cations	N-NH_4^+	0.06–0.77			
		[µmol L^{-1}]	[mg L^{-1}]	[µEq L^{-1}]	
	Na^+	0.8–486	0.12–9.76	2–2000	[34, 38, 65–68]
	NH_4^+	0.4–5.4	<LOD-0.11	–	[38, 40, 49, 67]
	K^+	0.03–11.5	<LOD-5.56	<LOD-96	[34, 38, 49, 65–68]
	Mg^{2+}	0.3–47.9	<LOD-1.378	<LOD-200	[34, 38, 49, 65–68]
	Ca^{2+}	0.6–15.2	0.04–2.17	<LOD-110	[34, 38, 49, 65–68]

Determined compound(s)/parameters		Identified level/range	References
Si-Si(OH)$_4$		[mg L^{-1}]	
		0.03–0.13	[65]
SiO_2		0.0	[34]
Si		<LOD-1.5 [μmol L^{-1}]	[68]
Metals	Al$_{total}$	3.78-117 [μg L^{-1}]	[65]
	Hg$_{total}$	[ng L^{-1}]	
		<LOD-59.9	[67, 69]
	Hg$_{reactive}$	2.2–45.3	[69]
	MMHg	3–43 [pg L^{-1}]	[67]
S		0.17–0.88 [mg L^{-1}]	[66]
MSA (methyl-sulfonic acid)		[μmol L^{-1}]	
		<LOD-1.56	[67]
Glut (glutaric acid)		<LOD-0.07	
∑PCB$_9$		[pg L^{-1}]	
		116–2000	[70, 71]
α-HCH		<LOD-47.6	
γ-HCH		186–3090	
∑DDT		0.391–59.5	
HCB		3.10–35.3	
∑PCN		59.0–1100	[71]
PFOA		89.5–590.8	[49]
PFOS		18.6–133.2	

Table 5a. Literature data on snow samples collected in the glacial catchments of Svalbard.

the pHs closest to neutral and lowest electrical conductivities, and also in terms of inorganic ions their concentration range was smaller than experienced in snow samples (**Table 5b** and **5a**). This reflects the effects of snow accumulation on inorganic chemicals, which are readily removed in meltwater (**Table 5c**) and therefore less of them remains in glacial ice. Conversely, the POPs found in ice were usually occurring at higher concentration than in snow, showing their historical deposition was higher, but also perhaps the ability of the accumulating snowpack to retain them better. An environmental concern are also the concentrations of heavy metals experienced in glacial ice (**Table 5b**), which additionally demonstrate the possibility that glaciers store pollutants of various types.

In **Table 6** we additionally provide the data on other abiotic media except frozen and liquid water, i.e. soil and cryoconite sediment. For cryoconite, it is noteworthy that it may contain

Determined compound(s)/parameters		Identified level/range		References
pH [−]		5.65–7.03		[88]
SEC [μS cm^{-1}]		4.50–21.2		
Anions		[μmol L^{-1}]	[mg L^{-1}]	
	Cl^-	328	<LOD-1.12	[38, 49]
	NO_3^-	1.5	<LOD-0.10	
	SO_4^{2-}	19.8	<LOD-0.27	
∑anions	F^-, Cl^-, NO_2^-, Br^-, NO_3^-, PO_4^{3-}, SO_4^{2-}	[meq L^{-1}]		
		0.022–0.236		[88]
Cations		[μmol L^{-1}]	[mg L^{-1}]	
	Na^+	199	<LOD-0.7	[38, 49]
	NH_4^+	0.4	<LOD-0.08	
	K^+	7.6	<LOD-0.09	
	Mg^{2+}	26.2	<LOD-0.19	
	Ca^{2+}	28.5	0.03–0.75	
∑cations	Na^+, NH_4^+, Li^+, K^+, Mg^{2+}, Ca^{2+}	[meq L^{-1}]		[88]
		0.015–0.279		
Metals		[μg kg^{-1}]	[μg L^{-1}]	[55, 72, 88] [55, 72, 88]
	Zn	43.75	1–40.91	
	Mn	42.75	0.22–5.20	
	Cu	11.25	0.27–3.25	
	Fe	2552.50	0.10–17.20	
	Ni	7.25	0.13–2.34	
	Cr	19.25	<LOD-0.16	
	Pb	16.75	0.02–0.45	
	Cd	4.50	<LOD-0.10	
	Co	1.50	<LOD	
	Be	–	<LOD-0.02	
	B	–	<LOD-2.31	
	Al	–	<LOD-2.85	
	Se	–	<LOD-0.15	
	Rb	–	<LOD-0.30	
	Sr	–	0.51–3.89	
	Ba	–	0.30–3.14	
	U	–	<LOD-0.02	
PFOA		[pg L^{-1}]		[49]
		13.5–45.9		

Determined compound(s)/parameters		Identified level/range		References
PFOS		<LOQ-13.5		
DOC		[mg L^{-1}]		[88]
		<LOD-0.566		
Pesticides		[pg L^{-1}]	[pg cm^{-2} yr^{-1}]	[73, 74]
	Aldrin	69,000	30,000	
	Dieldrin	7500	54.7	
	Endosulfan (α, β)	10,700–19,700	2.8–6.8	
	Endrin	–	16.3	
	Endrin-aldehyde	13,600	–	
	Endrin-ketone	–	13.6	
	Heptachlor	6500	470	
	Heptachlor epoxide	32,800	1580	
	HCH (α, γ)	1100–7700	295–369	
	Methoxychlor	4700	19.6	
	Chlorpyrifos	16,200	809	
	Dacthal	300	12.7	
	Diazinon	20,500	1410	
	Dimethoate	87000	598	
	Disulfoton	6500	447	
	Imidan	44,100	3030	
	Methylparathion	7400	357	
	Terbufos	11,100	530	
	Alachlor	1200	57	
	Desethyl-atrazine	2100	144	
	Metolachlor	9300	450	
	Pendimethalin	18,600	890	
	Chlordane (α, γ)	–	13.39–18.3	
	DDD(o.p')	–	11.5	
	DDE (p.p')	–	1.14	
	DDT (L) (p.p')	–	2.93	
	Endosulfan sulphate	–	2.81	
	Metribuzin	–	1.05	
	Nonachlor (trans, cis)	–	2.28–5.03	
	Trifluralin	–	2.32	

Table 5b. Literature data on chemical concentrations in ice samples from Svalbard glaciers.

Determined compound(s)/parameters		Identified level/range			References
pH		4.95–9.74			[34–38, 65]
EC		[µS cm^{-3}] 84.00–188.5			[35]
Anions	Cl^-	[µmol L^{-1}]	[mg L^{-1}]	[µEq L^{-1}]	[12, 34, 35, 37, 38, 65, 76]
		58–464	0.41–36	4–991	
	NO_3^-	0.1–34.7	0.4–2.2	0.56–9.0	[34, 35, 38, 40, 76]
	PO_4^{3-}	–	<LOD-1.0	–	[12]
	SO_4^{2-}	12–217.8	0.73–920.0	1–27,400	[12, 34–38, 65, 75, 76]
HCO_3^-		[µmol L^{-1}]	[µEq L^{-1}]		[12, 34–38, 65, 75]
		4.65–3198	<LOD-7600		
N-NO_2^-		[mg L^{-1}]			[12]
		<LOD-4.90			
N-NO_3^-		0.01–50.70			[12, 65]
N-NH_4^+		<LOD-19.50			[12, 65]
Cations	Na^+	[µmol L^{-1}]	[mg L^{-1}]	[µEq L^{-1}]	[12, 34–38, 65, 76]
		79–513	0.39–35.1	4–833	
	NH_4^+	0.1–6	–	–	[38, 40]
	K^+	8–26	0.22–5.5	<LOD-37	[12, 34, 35, 37, 38, 65, 76]
	Mg^{2+}	92–633	0.18–75.3	<LOD-12,300	[12, 34–38, 65, 75, 76]
	Ca^{2+}	249–1072	0.54–33,300	9–18,700	[12, 34–38, 65, 75, 76]
Metals	Al_{total}	[mg L^{-1}]			[65, 72]
		1.9–275.0			
	Fe	<0.010–0.300			
	Mn	<0.050			
	Zn	<0.001–0.010			
	Pb	<0.001–0.010			
	Cu	<0.001			
	Cd	<0.001			
Si		[mg L^{-1}]			[76]
		0.46–2.31			
Si-$Si(OH)_4$		0.01–0.63			[65]
SiO_2		[µmol L^{-1}]	[mg L^{-1}]	[µEq L^{-1}]	[34–37]
		2-22	0.120–0.780	2–34	
DIC		15.3–851.3			[76]
DOC		0.31–2.17			[38, 65]
		[µmol L^{-1}]			

Determined compound(s)/parameters	Identified level/range	References
	165–426	
DON	<7–27	[38]
DIN	<LOD-132.5 [μg N L^{-1}]	[77]
PFOA	[pg L^{-1}]	[49]
	95.7–639	
PFOS	<LOQ-967	
	–	
TIN	<LOD-18.2	
TN	2200–3800	

Table 5c. Literature overview of chemical concentrations in surface water samples from the glacial catchments of Svalbard.

marked amounts of both harmful heavy metals and life-supporting nutrients. In respect to soils, it can be highlighted that their ionic components may be at lower concentrations than those encountered in the riverine waters flowing out of glacial catchments, especially the fast-flowing, sediment-rich proglacial rivers (**Table 5c**).

Type of abiotic sample	Determined compound(s)/parameters		Identified level/range		References
Soil	pH		7.38–8.79		[34, 65]
	Anions	Cl^-	[μEq L^{-1}]		[34]
			120		
		NO_3^-	19		
		SO_4^{2-}	240		
	HCO_3^-		4100		[34]
	Cations	Na^+	[mmol kg^{-1}]	[μEq L^{-1}]	[34, 65]
			0.12–1.72	180	
		K^+	1.74–4.04	31	
		Mg^{2+}	4.11–23.4	1700	
		Ca^{2+}	63.2–528	2700	
	Organic carbon		[%]	[65]	
			1.28–6.05		
	N		0.04–0.16		[65]
	SiO_2		3.3 [mg L^{-1}]		[34]
Cryoconite (sediment)	Metals	Fe	[g kg^{-1}]		[72]
			31.9		

Type of abiotic sample	Determined compound(s)/parameters	Identified level/range	References
	Mn	0.11	
	Zn	0.08	
	Pb	0.19	
	Cu	–	
	Cd	–	
	DIN	[μg N g^{-1}]	[77]
		–	
	TIN	<LOD-18.2	
	TN	2200–3800	

Table 6. Literature data on sediment samples (soil, cryoconite) collected in the glacial catchments of Svalbard.

5. Summary

The results of research presented in the reviewed literature do not answer all questions arising in the context of the current global warming. The role of glaciers as contributors to the sea level rise is widely discussed by scientists, and extensively described in the latest IPCC report. However, the role of the changing glaciers and glacial waters, particularly for the biota of the polar environment, frequently does not receive enough attention. Considering the presence of contaminants such as POPs in many abiotic elements of the glacial catchment (e.g., snow, ice, surface water), it seems necessary to ask questions about the way and pace of the release of these highly toxic contaminants from the rapidly melting Arctic glaciers, as well as about the potential impact of those on the polar wildlife. Further research should address these questions, in order to help protect the highly sensitive environment of this area. Especially, a more detailed approach to the transport, deposition, and redistribution or transformation of pollutants in the glacial catchments is required, as opposed to the focus on the presence of pollutants in the environment only. However, without a stronger basis in chemical monitoring, there is frequently too little data to draw more global conclusions about the fate of chemicals in Svalbard glaciers.

Acknowledgements

K. Koziol acknowledges National Science Centre in Poland (NCN) funding, grant no. 2012/05/N/ST10/02848. The publication has been (partially) financed from the funds of the Leading National Research Centre (KNOW) received by the Centre for Polar Studies of the University of Silesia, Poland.

List of acronyms

AAS	Atomic Absorption Spectroscopy
ACIA	Arctic Climate Impact Assessment
AFS	Atomic Fluorescence Spectrometry
ASTER	Advanced Spaceborne Thermal Emission and reflection Radiometer
CVAFS	Cold Vapour Atomic Fluorescence Spectroscopy
DDD	Dichlorodiphenyldichloroethane
DDE	Dichlorodiphenyldichloroethylene
DDT	Dichlorodiphenyltrichloroethane
DIC	Dissolved Inorganic Carbon
DIN	Dissolved Inorganic Nitrogen
DOC	Dissolved Organic Carbon
DON	Dissolved Organic Nitrogen
EC	Electrochemical Conductivity
ECV	Essential Climate Variables
ERC	European Research Council
ESF	European Science Foundation
FAO	Food and Agriculture Organization
FIA	Flow Injection Analysis
GC-ECD	Gas Chromatography with Electron Capture Detection
GC-LRMS-ECNI	Gas Chromatography Coupled to Low Resolution Mass Spectrometry in Electron Capture Negative Ionization Mode
GC-LRMS-EI	Gas Chromatography Coupled to Mass Spectrometry with Low Resolution in Electron Ionization Mode
GC-MS-EI-SIM	Gas Chromatography Coupled to Mass Spectrometry in Electronic Ionization Mode with Selected-Ion Monitoring
GLIMS	Global Land Ice Measurements from Space
GLS	Global Land Surface
GOA	Global Ocean Area
GTN	Global Terrestrial Network

GTN-G	Global Terrestrial Network for Glaciers
GTOS	Global Terrestrial Observing System
HCB	Hexachlorobenzene
HCH	Hexachlorocyclohexane
HRGC-LRMS	High-Resolution Gas Chromatography Coupled with Low Resolution Mass Spectrometry
IC	Ion Chromatography
ICP-OES	Inductively Coupled Plasma Optical Emission Spectrometry
ICP-SFMS	Inductively Coupled Plasma-Sector Field Mass Spectrometry
ICP-QMS	Inductively Coupled Plasma-Quadrupole Mass Spectrometry
ICSU	International Council for Science
IPCC	Intergovernmental Panel on Climate Change
LC-MS/MS	Liquid Chromatography with Tandem Mass Spectrometry
LOD	Limit Of Detection
LOQ	Limit Of Quantitation
LRAT	Long Range Atmospheric Transport
NSIDC	National Snow and Ice Data Center
PCB	Polychlorinated Biphenyls
PCN	Polychlorinated Naphthalene
PFOA	Perfluorooctanoate
PFOS	Perfluorooctane sulfonate
POP	Persistent Organic Pollutants
PSFG	Permanent Service on the Fluctuations of Glaciers
RCN	Research Council of Norway
TIN	Total Inorganic Nitrogen
TN	Total Nitrogen;
TOC	Total Organic Carbon;
TON	Total Organic Nitrogen;
TOPC	Terrestrial Observation Panel for Climate;

TTS/WGI	Temporal Technical Secretariat for the World Glacier Inventory;
UNEP	United Nations Environment Programme;
UNESCO	United Nations Educational, Scientific and Cultural Organization;
WGMS	World Glacier Monitoring Service;
WMO	World Meteorological Organization.

Author details

Sara Lehmann-Konera[1], Marek Ruman[2,3]*, Krystyna Kozioł[1], Grzegorz Gajek[4] and Żaneta Polkowska[1]*

*Address all correspondence to: marek.ruman@us.edu.pl; zanpolko@pg.gda.pl

1 Department of Analytical Chemistry, Faculty of Chemistry, Gdansk University of Technology, Gdansk, Poland

2 Faculty of Earth Sciences, University of Silesia, Sosnowiec, Poland

3 Centre for Polar Studies, Leading National Research Centre (KNOW), Sosnowiec, Poland

4 Department of Geology and Lithosphere Protection, Faculty of Earth Sciences and Spatial Management, Maria Curie-Skłodowska University in Lublin, Lublin, Poland

References

[1] Bogdal C, Schmid P, Zennegg M, Anselmetti FS, Scheringer M, Hungerbühler K. Blast from the past: Melting glaciers as a relevant source for persistent organic pollutants. Environmental Science & Technology. 2009;**43**:8173–8177

[2] Grannas AM, Bogdal C, Hageman KJ, Halsall C, Harner T, Hung H, Kallenborn R, Klán P, Klánová J, Macdonald RW, Meyer T, Wania F. The role of the global cryosphere in the fate of organic contaminants. Atmospheric Chemistry and Physics. 2013;**13**:3271–3305

[3] Hodson AJ. Understanding the dynamics of black carbon and associated contaminants in glacial systems. WIREs Water. 2014;**1**:141–149. DOI: 10.1002/wat2.1016

[4] ACIA. Arctic Climate Impact Assesment. Cambridge University Press. Cambridge: 2005. p. 1042. http://www.acia.uaf.edu (19.07.2014): 183–242

[5] IPCC 2013, Vaughan DG, Comiso JC, Allison I, Carrasco J, Kaser G, Mote P, et al. Observations: cryosphere. In: Stocker TF, Qin D, Plattner GK, Tignor M, Allen SK, Boschung J, et al, editors. Climate Change 2013: The Physical Science Basis. Contribution of working group I to the Fifth Assessment Report of the Intergovernmental Panel on Climate

Change. Cambridge, United Kingdom and New York, NY, USA: Cambridge University Press; 2013

[6] Jansson P, Hock R, Schneider T. The concept of glacier storage: A review. Journal of Hydrology. 2003;**282**(1):116–129

[7] Błaszczyk M, Jania JA, Hagen JO. Tidewater glaciers of Svalbard: Recent changes and estimates of calving fluxes. Polish Polar Research. 2009;**30**(2):85–142

[8] Nuth C, Moholdt G, Kohler J, Hagen JO and Kääb A. Svalbard glacier elevation changes and contribution to sea level rise. Journal of Geophysical Research. 2010;**115**:F01008. DOI: 10.1029/2008JF001223

[9] Ruman M, Kozak K, Lehmann S, Kozioł K, Polkowska Ż. Pollutants present in different components of the Svalbard archipelago environment. Ecological Chemistry and Engineering S. 2012;**19**(4):571–584

[10] Kozak K, Polkowska Ż, Ruman M, Kozioł K, Namieśnik J. Analytical studies on the environmental state of the Svalbard Archipelago provide a critical source of information about anthropogenic global impact. TrAC Trends in Analytical Chemistry. 2013;**50**:107–126

[11] Jiao L, Zheng GJ, Minh TB, Richardson B, Chen L, Zhang Y, et al. Persistent toxic substances in remote lake and coastal sediments from Svalbard, Norwegian Arctic: Levels, sources and fluxes. Environmental Pollution. 2009;**157**:1342–1351

[12] Dragon K, Marciniak M. Chemical composition of groundwater and surface water in the Arctic environment (Petuniabukta region, central Spitsbergen). Journal of Hydrology. 2010;**386**(1):160–172

[13] Brown GH. Glacier meltwater hydrochemistry. Applied Geochemistry. 2002;**17**(7):855–883

[14] Cuffey K, Paterson WSB. The Physics of Glaciers. 4th ed. Academic Press; San Diego, United States: 2010

[15] Irvine Flynn TDL, Hodson AJ, Mooram BJ, Vatne G, Hubbard AL. Polythermal glacier hydrology: A review. Reviews of Geophysics. 2011;**49**(4):RG4002. DOI: 10.1029/2010RG000350

[16] Jania J. Glacjologia. Nauka o Lodowcach. Warszawa: PWN. 1993. p. 358

[17] Sugden DE, John BS. Glaciers and Landscape. A Geomorphological Approach. London: Edward Arnold. 1976. p. 376

[18] Głowacki P. Rola procesów fizyczno-chemicznych w kształtowaniu struktury wewnętrznej i obiegu masy lodowców Spitsbergenu. Publications of the Institute of Geophysics. Polish Academy of Sciences. 2007;**M-30**(400):147

[19] Kääb A, Huggel C, Paul F, Wessels R, Raup B, Kieffer H, Kargel J. Glacier monitoring from ASTER imagery: Accuracy and applications. In Proceedings of EARSeL-LISSIG-Workshop Observing our Cryosphere from Space. 2002;**2**: 43–53

[20] Roer I. Global glacier changes: Facts and figures. In: Zemp M, van Woerden J, editors. UNEP/Earthprint; 2008

[21] GLIMS Information from: http://www.glims.org/[date:19.08.2014]

[22] Zemp M. The monitoring of glaciers at local, mountain, and global scale. Geographisches Institut der Universität Zürich. 2012;**65**

[23] Kargel JS. Global land ice measurements from space. In: Leonard GJ, Bishop MP, Kääb A, Raup BH, editors. Heidelberg: Springer; 2013. ISBN 978-3-540-79818-1 (eBook)

[24] GTN-G Information. Available from: http://www.gtn-g.org/ [Accessed date August 19, 2014]

[25] UIO Information. Available from: http://www.mn.uio.no [Accessed date August 19, 2014]

[26] ICE2SEA Information. Available from: http://www.ice2sea.eu [Accessed date August 19, 2014]

[27] SVALGLAC Information. Available from: http://svalglac.eu [Accessed date August 19, 2014]

[28] GLACIODYN Information. Available from: http://www.unis.no/ [Accessed date August 19, 2014]

[29] Sharp M, Wolken G, Geail ML, Burgess D, Cogley JG, Arendt A, Wouters B. Mountain Glaciers and Ice Caps (Outside Greenland)[in Arctic Report Card 2013], ftp://ftp.oar.noaa.gov/arctic/documents/ArcticReportCard_full_report2013.pdf [date: 19.08.2014], p. 108.

[30] Hagen JO, Liestøl O, Roland E, Jørgensen T. Glacier atlas of Svalbard and Jan Mayen. Norsk Polarinstitutt Meddelelser; 1993. p. 129

[31] Ziaja W, Skiba S. [ed.] Structure and functioning of the Sorkappland natural environment [in Polish: Struktura i funkcjonowanie środowiska przyrodniczego Sorkapplandu]. Kraków: UJ; 2002.

[32] Hagen JO, Kohler J, Melvold K, Winther JG. Glaciers in Svalbard: Mass balance, runoff and freshwater flux. Polar Research. 2003;**22**(2):145–159

[33] Głowacki P. The mass balance of Hans glacier in the light of cryochemical investigation. In: P Głowacki, editor. Pol Polar Stud 1997. 24th Polar Symposium, Warszawa, 1997. Warszawa: Institute of Geophysics of the Polish Academy of Sciences; 1997. pp. 75–79

[34] Hodson A, Tranter M, Gurnell A, Clark M, Hagen JO. The hydrochemistry of Bayelva, a high Arctic proglacial stream in Svalbard. Journal of Hydrology. 2002;**257**:91–114

[35] Krawczyk WE, Lefauconnier B, Pettersson LE. Chemical denudation rates in the Bayelva Catchment, Svalbard, in the Fall of 2000. Physics and Chemistry of the Earth Parts A/B/C. 2003;**28**(28):1257–1271

[36] Nowak A, Hodson A. Changes in meltwater chemistry over a 20-year period following a thermal regime switch from polythermal to cold-based glaciation at Austre Brøggerbreen, Svalbard. Polar Research. 2014;**33**: 1-19. Article number: 22779. DOI: 10.3402/polar.v33.22779

[37] Stachnik Ł, Majchrowska E, Yde JC, Nawrot AP, Cichała-Kamrowska K, Ignatiuk D, Piechota A. Chemical denudation and the role of sulfide oxidation at Werenskioldbreen, Svalbard. Journal of Hydrology. 2016;**538**:177–193

[38] Tye AM, Heaton THE. Chemical and isotopic characteristics of weathering and nitrogen release in non-glacial drainage waters on Arctic tundra. Geochimica et Cosmochimica Acta. 2007;**71**:4188–4205

[39] Stachnik Ł, Yde JC, Kondracka M, Ignatiuk D, Grzesik M. Glacier naled evolution and relation to the subglacial drainage system based on water chemistry and GPR surveys (Werenskioldbreen, SW Svalbard). Annals of Glaciology. 2016;**57**:1–12

[40] Wynn PM, Hodson AJ, Heaton THE, Chenery SR. Nitrate production beneath a high Arctic glacier, Svalbard. Chemical Geology. 2007;**244**:88–102

[41] AMAP 2016. AMAP Assessment 2015: Temporal Trends in Persistent Organic Pollutants in the Arctic. Oslo; pp. i-71

[42] Wang-Andersen G, Utne Skaare J, Prestrud P, Steinnes E. Levels and congener pattern of PCBs in arctic fox (*Alopex lagopus*) in Svalbard. Environmental Pollution. 1993;**82**(3):269–275

[43] Bernhoft A, Wiig Ø, Skaare JU. Organochlorines in polar bears (*Ursus maritimus*) at Svalbard. Environmental Pollution. 1997;**95**(2):159–175

[44] Kleivane L, Severinsen T, Skaare JU. Biological transport and mammal to mammal transfer of organochlorines in Arctic fauna. Marine Environmental Research. 2000;**49**(4):343–357

[45] Henriksen EO, Wiig Ø, Skaare JU, Gabrielsen GW, Derocher AE. Monitoring PCBs in polar bears: Lessons learned from Svalbard. Journal of Environmental Monitoring. 2001;**3** (5):493–498

[46] Fuglei E, Bustnes JO, Hop H, Mørk T, Björnfoth H, Van Bavel B. Environmental contaminants in arctic foxes (*Alopex lagopus*) in Svalbard: Relationships with feeding ecology and body condition. Environmental Pollution. 2007;**146**(1):128–138

[47] Routti H, Letcher RJ, Chu S, Van Bavel B, Gabrielsen GW. Polybrominated diphenyl ethers and their hydroxylated analogues in ringed seals (*Phoca hispida*) from Svalbard and the Baltic Sea. Environmental Science & Technology. 2009;**43**(10):3494–3499

[48] Letcher RJ, Bustnes JO, Dietz R, Jenssen BM, Jørgensen EH, et al. Exposure and effects assessment of persistent organohalogen contaminants in arctic wildlife and fish. Science of the Total Environment. 2010;**408**(15):2995–3043

[49] Kwok KY, Yamazaki E, Yamashita N, Sachi Taniyasu S, Murphy MB. Transport of perfluoroalkyl substances (PFAS) from an arctic glacier to downstreamlocations: Implications for sources. Science of the Total Environment. 2013;**447**:46–55

[50] Hallanger IG, Ruus A, Herzke D, Warner NA, Evenset A, et al. Influence of season, location, and feeding strategy on bioaccumulation of halogenated organic contaminants in Arctic marine zooplankton. Environmental Toxicology and Chemistry. 2011;**30**(1):77–87

[51] Vallack HW, Bakker DJ, Brandt I, Broström-Lundén E, Brouwer A. Controlling persistent organic pollutants–What next? Environmental Toxicology and Pharmacology. 1998;**6**(3): 143 175

[52] Lohmann R, Breivik K, Dachs J, Muir D. Global fate of POPs: Current and future research directions. Environmental Pollution. 2007;**150**(1):150–165

[53] Macdonald RW, Barrie LA, Bidleman TF, Diamond ML, Gregor DJ. Contaminants in the Canadian Arctic: 5 years of progress in understanding sources, occurrence and pathways. Science of the Total Environment. 2000;**254**(2):93–234

[54] Wania F, Mackay D. The evolution of mass balance models of persistent organic pollutant fate in the environment. Environmental Pollution. 1999;**100**(1):223–240

[55] Drbal K, Elster J, Komarek J. Heavy metals in water, ice and biological material from Spitsbergen, Svalbard. Polar Research. 1992;**11**(2):99–101

[56] Wang Z, Ma X, Na G, Lin Z, Ding Q, Yao Z. Correlations between physicochemical properties of PAHs and their distribution in soil, moss and reindeer dung at Ny-Ålesund of the Arctic. Environmental Pollution. 2009;**157**(11):3132–3136

[57] Hoekstra PF, Braune BM, O'Hara TM, Elkin B, Solomon KR, Muir DCG. Organochlorine contaminant and stable isotope profiles in Arctic fox (*Alopex lagopus*) from the Alaskan and Canadian Arctic. Environmental Pollution. 2003;**122**(3):423–433

[58] Gabrielsen GW. Levels and effects of persistent organic pollutants in arctic animals. In Arctic Alpine Ecosystems and People in a Changing Environment. Berlin Heidelberg; Springer; 2007. pp. 377–412

[59] Bang K, Jenssen BM, Lydersen C, Skåre JU. Organochlorine burdens in blood of ringed and bearded seals from north-western Svalbard. Chemosphere. 2001;**44**(2):193–203

[60] Weber J, Halsall CJ, Muir D, Teixeira C, Small J, et al. Endosulfan, a global pesticide: A review of its fate in the environment and occurrence in the Arctic. Science of the Total Environment. 2010;**408**(15):2966–2984

[61] Bidleman TF, Helm PA, Braune BM, Gabrielsen GW. Polychlorinated naphthalenes in polar environments—A review. Science of the Total Environment. 2010;**408**(15):2919–2935

[62] Wong F, Jantunen LM, Pucko M, Papakyriakou T, Staebler RM, Stern GA, Bidleman TF. Air-water exchange of anthropogenic and natural organohalogens on International Polar Year (IPY) expeditions in the Canadian Arctic. Environmental Science & Technology. 2010;**45**(3):876–881

[63] Kallenborn R, Christensen G, Evenset A, Schlabach M, Stohl A. Atmospheric transport of persistent organic pollutants (POPs) to Bjørnøya (Bear island). Journal of Environmental Monitoring. 2007;**9**(10):1082–1091

[64] Hung H, Kallenborn R, Breivik K, Su Y, Brorström-Lundén E, et al. Atmospheric monitoring of organic pollutants in the Arctic under the Arctic Monitoring and Assessment Programme (AMAP): 1993–2006. Science of the Total Environment. 2010;**408**(15):2854–2873

[65] Stutter MI, Billett MF. Biogeochemical controls on streamwater and soil solution chemistry in a High Arctic environment. Geoderma. 2003;**113**(1):127–146

[66] De Caritat P, Hall G, Gìslason S, Belsey W, Braun M. Chemical composition of arctic snow: Concentration levels and regional distribution of major elements. Science of the Total Environment. 2005;**336**(1):183–199

[67] Larose C, Dommergue A, De Angelis M, Cossa D, Averty B, et al. Springtime changes in snow chemistry lead to new insights into mercury methylation in the Arctic. Geochimica et Cosmochimica Acta. 2010;**74**(22):6263–6275

[68] Hodgkins R, Tranter M. Solute in High Arctic glacier snow cover and its impact on run-off chemistry. Annals of Glaciology. 1998;**26**:156–160

[69] Ferraria CP, Padova C, Faïn X, Gauchard PA, Dommergue A, et al. Atmospheric mercury depletion event study in Ny-Alesund (Svalbard) in spring 2005. Deposition and transformation of Hg in surface snow during springtime. Science of the Total Environment. 2008;**397**(1):167–177

[70] Herbert BMJ, Halsall CJ, Villa S, Jones K C, Kallenborn R. Rapid changes in PCB and OC pesticide concentrations in Arctic snow. Environmental Science & Technology. 2005a;**39** (9):2998–3005

[71] Herbert BMJ, Halsall CJ, Villa S, Fitzpatrick L, Jones KC, et al. Polychlorinated naphthalenes in air and snow in the Norwegian Arctic: A local source or an Eastern Arctic phenomenon? Science of the Total Environment. 2005b;**342**(1):145–160

[72] Chmiel S, Reszka M, Rysiak A. Heavy metals and radioactivity in environmental samples of the Scott Glacier region on Spitsbergen in summer 2005. Quaestiones Geographicae 28A/1. Poznań: Adam Mickiewicz University Press; 2009. pp. 23–29. ISBN 978-83-232-2128-9

[73] Hermanson MH, Isaksson E, Teixeira C, Muir DC, Compher KM, et al. Current-use and legacy pesticide history in the Austfonna ice cap, Svalbard, Norway. Environmental Science & Technology. 2005;**39**(21):8163–8169

[74] Ruggirello RM, Hermanson MH, Isaksson E, Teixeira C, Forsström S, et al. Current use and legacy pesticide deposition to ice caps on Svalbard, Norway. Journal of Geophysical Research Atmospheres. 2010;**115**(D18):1–11. Article number: D18308

[75] Wadham JL, Cooper RJ, Tranter M, Bottrell S. Evidence for widespread anoxia in the proglacial zone of an Arctic glacier. Chemical Geology. 2007;**243**(1):1–15

[76] Rutter N, Hodson A, Irvine-Fynn T, Solås MK. Hydrology and hydrochemistry of a deglaciating high-Arctic catchment, Svalbard. Journal of Hydrology. 2011;**410**(1):39–50

[77] Telling J, Anesio AM, Tranter M, Irvine-Fynn T, Hodson A, et al. Nitrogen fixation on Arctic glaciers, Svalbard. Journal of Geophysical Research Biogeosciences. 2011;**116**:1–8;Article number: G03039

[78] Pulina M, Burzyk M. Fuglebekken catchment [in Polish]. In: Glaciological workshops 2004; Polish Geomorphologists Association, Sosnowiec-Poznań-Longyearbyen. 2004;**VI**:58–62

[79] Stibal M, Tranter M. Laboratory investigation of inorganic carbon uptake by cryoconite debris from Werenskioldbreen, Svalbard. Journal of Geophysical Research. 2007;**112**:9; G04S33. DOI: 10.1029/2007JG000429

[80] Taurisano A, Schuler TV, Hagen JO, Eiken T, Loe E, Melvold K, Kohler J. The distribution of snow accumulation across the Austfonna ice cap, Svalbard: Direct measurements and modelling. Polar Research. 2007;**26**(1):7–13

[81] Kääb A. Glacier volume changes using ASTER satellite stereo and ICESat GLAS laser altimetry. A test study on edgeoya, eastern svalbard. In: Paper presented at IEEE International Geoscience and Remote Sensing Symposium (IGARSS); Jul 23–27; Barcelona, Spain. IEEE-Instrumentation Electrical Electronics Engineers Inc; 2008

[82] Rolstad C, Norland R. Ground-based interferometric radar for velocity and calving-rate measurements of the tidewater glacier at Kronebreen, Svalbard. Annals of Glaciology. 2009;**50**:47–54

[83] Sund M, Eiken T, Hagen JO, Kääb A. Svalbard surge dynamics derived from geometric changes. Annals of Glaciology. 2009;**50**(52):50–60

[84] Moholdt G, Hagen JO, Eiken T, Schuler TV. Geometric changes and mass balance of the Austfonna ice cap, Svalbard. The Cryosphere. 2010;**4**:21–34. DOI: 10.5194/tc-4-21-2010

[85] Schuler TV, Loe E, Taurisano A, Eiken T, Hagen JO, Kohler J. Calibrating a surface mass-balance model for Austfonna ice cap, Svalbard. Annals of Glaciology. 2007;**46**(1):241–248

[86] Norheim G, Kjos-Hanssen B. Persistent chlorinated hydrocarbons and mercury in birds caught off the west coast of Spitsbergen. Environmental Pollution Series A. 1984;**33**(2):143–152

[87] Vieweg I, Hop H, Brey T, Huber S, Ambrose Jr WG, Locke VWL, Gabrielsen GW. Persistent organic pollutants in four bivalve species from Svalbard waters. Environmental Pollution. 2012;**161**:134–142

[88] Lehmann S, Gajek G, Chmiel S, Polkowska Ż. Do morphometric parameters and geological conditions determine chemistry of glacier surface ice? Spatial distribution of contaminants present in the surface ice of Spitsbergen glaciers (European Arctic). Environmental Science and Pollution Research. 2016;**23**(23):23385–23405

Chapter 2

A Review on the Little Ice Age and Factors to Glacier Changes in the Tian Shan, Central Asia

Yanan Li, Xiaoyu Lu and Yingkui Li

Additional information is available at the end of the chapter

http://dx.doi.org/10.5772/intechopen.70044

Abstract

Mountain glaciers are a reliable and unequivocal indicator of climate change due to their sensitive response to changes in temperature and precipitation. The importance of mountain glaciers is best reflected in regions with limited precipitation, such as arid and semi-arid central Asia. High concentration of glaciers and meltwater from the Tian Shan contribute considerably to the freshwater resource in Xinjiang (China), Kyrgyzstan and nearby countries. Documenting glacier distribution and research on glacier changes can provide insights and scientific support for water management in central Asia. As the most recent glacial event, the Little Ice Age (LIA, approximately AD 1300–1850) signifies the cold periods prior to the warming trend in the twentieth century. Here we present an overview of topics recently studied on the modern and LIA glaciers in the Tian Shan of the central Asia. With data sets of the Glacier Inventory of China and the presumed LIA glacial extents, we applied statistical models in a case study of the eastern Tian Shan to examine the impact of local topographic and geometric factors on glacier area changes. The findings of glacier size and elevation as key local factors are representative and consistent with other studies.

Keywords: glacier datasets, chronological dating, driving factors, random forest

1. Introduction

Extending the timescale of climate variations from the past century to the past millennium allows us to see the broader context of the unprecedented global warming today. Prior to the twentieth century warming, the cold extremes of the Little Ice Age (LIA) appeared to be remote in human's memory and often easily ignored. The term 'Little Ice Age' was first introduced by Matthes in 1939 in *Report of Committee on Glaciers* published on *Eos* [1]. Although his original term to describe 'an epoch of renewed but moderate glaciation that already has lasted about 4000 years' has been overtaken as '*neoglacial*', in this report, it was noted that '... the glacier-oscillations of the last few centuries have been among the greatest that have occurred during the 4000-year period...' [1]. The term has now been more formally and widely adopted to describe the period approximately from AD 1300 to 1850, characterized by lower temperature over most of the globe and growth of glaciers to a more advanced extent than that of prior and post the period [2].

Following the warm period of the 'Medieval Climatic Optimum', a dramatic series of glacier advances and retreats symbolized the cooling events of the LIA. However, despite the evidence of similar events from many places around the world, the occurrence of the LIA cannot be assumed synchronous in time and uniform across space. A large amount of evidence of the LIA was assembled initially in the regions like Europe, Greenland and the Arctic, around the North Atlantic, e.g. [3–8]. The expanded stages of mountain glaciers, cycles of excessive cold, severe droughts, hot summers, or unusual rainfall were captured in abundant historical documents such as artistic paintings, ship logs and agricultural records or diary from Mediterranean, alpine Europe and further north. For examples, see Refs. [4, 6, 9]. An increased variability of the climate, as well as other LIA-type events induced significant impacts on human society, as evidenced in existing historical documents. As noted in Ref. [10], the beginning of the LIA is marked by the heavy rains, severe winters and harvest fails in 1315–1319 widespread in England. But, it needs to be noted that firstly, the timing of cold conditions occurred differently from region to region, and secondly, throughout the span of the LIA, the climate was never monotonically cold or always favourable to glacial expansion but instead with sometimes disastrous shifts between warmth and coldness at centennial, decadal, or even annual scales [11]. Although the observational record is less available outside of the North Atlantic region, it is well studied that mountain glaciers advanced far beyond their modern limits in highlands of Asia, the Andes of South America, New Zealand, western North America and other ranges, e.g. [12–18]. Grove's book [2] provides a summary of the LIA-type events from all major regions over the world.

In this chapter, we focus on the Tian Shan range in central Asia, a less-studied region compared to other key mountains on the Earth, to synthesize (1) the critical regional settings of the Tian Shan glaciers; (2) current documentation and identification of modern glaciers and the LIA glacial events; and (3) the influence of climate and local factors to glacier change since the LIA. With LIA glacial chronologies established at a few sites across the Tian Shan, it makes the examination of the spatiotemporal pattern of the LIA glacial advances possible. The spatial variation of glacier changes reflects its response to climatic shifts as well as the local topography and geometry. What modifications in climate systems occurred here during the LIA? What local factors be attributed to explain the glacier variability? This chapter provides a holistic review of such questions based on current literature. At the end, we took an example of a dataset of 865 glaciers in the eastern Tian Shan to examine the significant local factors to glacier changes using random forest model, and we compared the results with previous findings.

2. Glaciers in the Tian Shan

As located in the arid and semi-arid environment, glaciers in high mountains in central Asia are sensitive indicators of climate change, as well as important freshwater storages in the region. Major glacier-covered mountains in central Asia include the Altay, the Urals, the Tian Shan, the Kunlun, the Karakoram and the Himalayas (**Figure 1**), and they represent a wide range of climatic and hydrological conditions. The Tian Shan is approximately located at 40.5°N to 43.5°N and 75.5°E to 94.8°E.

2.1. Geographical and climatic settings

'Tian Shan' means 'Heavenly Mountains' (Shan = Mountains) in Chinese. Formed by the collision of the Indian and Eurasian continental plates about 40 to 50 million years ago, the Tian Shan is a ~2500 km long mountain series stretching from the western boundary of Kyrgyzstan across most of the Xinjiang Uyghur Autonomous Region, China (**Figure 1**). It is the largest mountain chain in the world's midlatitude arid region. Its highest peak is Tumur (other names: Victory Peak, Jengish Chokusu, or Pobeda), 7439 m above sea level (a.s.l.), on the border between easternmost Kyrgyzstan, Kazakhstan and China. Bounded by the Taklimakan Desert to the south and the Gurbantonggut Desert to the north, the Tian Shan indispensably serves

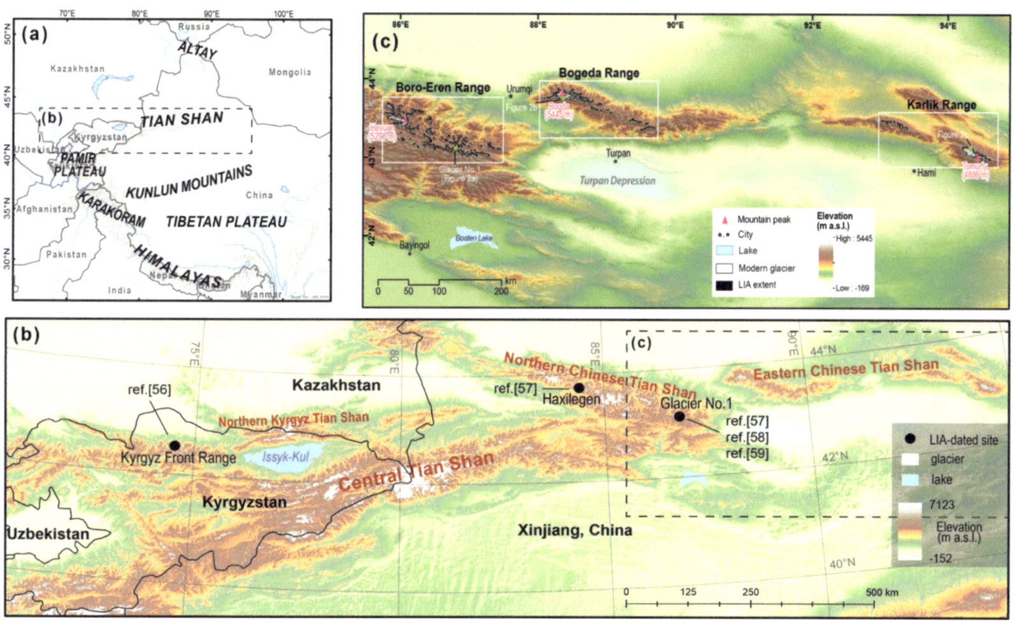

Figure 1. (a) Location of major mountains in central Asia and nearby area. (b) The Tian Shan with glacier data downloaded from the GLIMS Glacier Database (http://www.glims.org/download/); black dots represent the study sites with the LIA moraine identified by numerical dating methods [56–59]. (c) Study area of the eastern Chinese Tian Shan, with modern and LIA glacier data in three sub-regions.

as the source region of freshwater supply for the arid surroundings and nurtures abundant unique flora and fauna in its spectacular landscapes. Both the Western Tian Shan (Kyrgyzstan, Kazakhstan and Uzbekistan) and the Xinjiang Tian Shan (China) are listed as UNESCO World Heritage Site since 2013 and 2016, respectively (http://whc.unesco.org/en/list/).

The entire mountain system includes many individual mountain ranges shared across central Asia countries/regions. The western part of the system has relatively higher elevations, recognized as the Kyrgyz Tian Shan; whereas the Chinese Tian Shan (a.k.a. Xinjiang Tian Shan) is more characterized by smaller, lower ranges compared to the western section. The Lake Issyk-Kul (1608 m a.s.l.), situated in an intramountainous basin, dissects the Kyrgyz Tian Shan into a northern part (Zailiyskiy Alatau and Kungey Alatau Mountains) and a larger southern part (Terskey Alatau Mountains and Kok Shaal-tau Range) bordered with China. The altitude rises from about 700 m a.s.l. with steppe landscapes up to nearly 5000 m a.s.l. The central Tian Shan stretches from the eastern edge of Kyrgyzstan to China and is the highest mountain knot in the whole range where largest glaciers occupy. The Xinjiang Tian Shan in China covers the eastern portions of the Tian Shan, taking up about 1750 km, 2/3 of the total length. The Borohoro Range and the Tianger Range make up the northern branch of the Tian Shan.. Extending towards the east, there are the Bogeda Range located next to the regional capital city Urumqi and the Barkol-Karlik Range located to the easternmost end of the whole range (**Figure 1**).

The geographical location at the centre of the Earth's largest continent determines its typical continental, temperate climate in the Tian Shan. The confluence of major climate systems makes central Asia a transitional region, particularly sensitive to changes in the spatial pattern of climate systems. The overall continentality is characterized by sharp local differences. The elevation is the main factor influencing air temperature distribution rather than longitude or latitude. The contrasts in seasonal and diurnal temperature are high. The annual amplitude in monthly temperature of up to 38°C is common in the Kyrgyz Tian Shan, but regional differences exist [19]. Precipitation varies both longitudinally from west to east and from the northern slope to the southern slope. Such features are closely associated with two large-scale atmospheric circulations dominated in this area: midlatitude westerlies and the Siberian High. The moisture from the Atlantic Ocean and closed drainage basins such as the Aral, Black and Caspian seas is carried by the westerlies which delivers abundant precipitation to the western end of the mountains [20–22], while the orographic effect of the mountain barrier reduces it to the minimum amount at the eastern end. Sorg et al. [23] summarized that on the windward northwestern slopes, the annual precipitation can be up to 1500 to 2000 mm in high elevations, whereas to the east in the interior regions, it can be as low as 100 mm. In general, the Siberian anticyclonic circulation (high pressure) dominates the range in winter season with low temperatures helpful to maintain glaciers; after March, the westerlies associated with mid-latitude cyclones convey precipitation in form of snow to favour the growth of glaciers.

2.2. Status of current glaciers

A high concentration of glaciers in the Tian Shan is known as the "Water Tower of Central Asia" [23]. They play an important role in water cycle in this arid environment. The meltwater runoff into transboundary rivers, such as the Syr Darya River, the Ili River, the Kaidu River and the Urumqi River, contributes substantially to the freshwater resource in downstream ecosystems and for over 50 million populations in the central Asia region [23–25]. While melting glaciers in a first phase release an increasing amount of water, the undergoing reduction of glacier volume will eventually reduce water availability and induce potential political disputes on water allocation issue [26]. Documenting the distribution of present-day glaciers and studying the history of glacier changes in the Tian Shan are fundamental and critical for current and future development in this region.

Existing records of glaciers in the Tian Shan have been derived from aerial photographs, topographic maps and more common nowadays, satellite images, e.g. [27, 28]. The key role of remote sensing in glacier monitoring has been widely recognized, especially in areas like remote mountain ranges in the Tian Shan where the field-based glaciological survey is difficult to conduct. Two main parameters of glaciers obtained from remote-sensing data are the surface area and elevation, which have been commonly used to derive changes of glacial extents and thickness through time. Two international projects, World Glacier Monitoring Service (WGMS) [29] and Global Land-Ice Measurements from Space (GLIMS) [30], have been long devoted to create worldwide glacier inventories and to provide easy access to standard data for the public.

In the Tian Shan, glacier data have been collected both from the Kyrgyz side and the Chinese side. Aizen et al. [31] mapped glaciers in the Kyrgyz Tian Shan based on remote-sensing data and identified 7590 glaciers with a total area of 13,271 km^2 and an estimated 1840 km^3 volume of ice. The Chinese Academy of Sciences compiled the First Glacier Inventory of China (GIC) using topographic and aerial photographs acquired during the 1950s–1980s [32] and updated to the Second GIC in 2014 based on mostly Landsat images acquired between 2006 and 2010 [33]. This dataset is accessible via the Cold and Arid Regions Science Data Center at Lanzhou (http://card.westgis.ac.cn/), as well as through the GLIMS Glacier Database (http://glims.colorado.edu/glacierdata/). The GIC data followed the GLIMS guidelines and contain attributes such as glacier name, glacier identifier, drainage code, glacier area, absolute and relative accuracies, debris-covered area and many other parameters. According to the Second GIC, the total number of glaciers in the Chinese Tian Shan is 7927, and they cover an area of 7256 km^2 and an ice volume of 781 km^3. Such inventory data sets provide great convenience to know the status of present glaciers in the Tian Shan. Due to the fact that glaciers are changing constantly, successive inventories are anticipated to keep tracking the status of glaciers, under the potential of better available data and methods in future.

2.3. Evidence of LIA glaciers

Two distinct time periods, the 'Medieval Warm Period' and the 'LIA', represent unique but opposite climate conditions during the last millennium, and they set contemporary glacier changes into a long-term context [6]. Arguments and inconsistency exist on the accurate definition of the two climate episodes. Relatively abundant records in Europe and North America help identify such millennial or centennial glacier activities, but meanwhile, reflect a bias in the spatial representation when limited data are available in other key regions like central Asia. Until recent decades, studies on reconstructing past climate and glacier variations in the Tian Shan started to sprout, thanks to the development of techniques and better accessibility to the place.

In the Tian Shan, the terminal moraine characterized with no or little vegetation cover, massive piling of loose tills, sharply-crested ridge and a location a few hundred meters apart from glacier front is often believed as the mark of the LIA maximum extent [17, 34–36] (**Figure 2**). Numerical dating using proxy records is the approach for assigning the formation ages of such moraines and thus the timing of glacier fluctuations. These proxy materials include erratic boulders, buried wood, trees and lichens. Lichenometry is the dating method that derives moraine age from measuring the diameter of the largest/oldest lichen and uses the lichen's growth rate to infer the time of the moraine's formation. Although this method has been adopted since the 1950s, the difficulties such as species identification, establishing accurate growth curve and lack of knowledge of its colonization time on boulders creates many uncertainties and limits its use, e.g. [37–40]. Radiocarbon dating of organic matters, such as buried tree trunks that have been incorporated in moraine sediments or transported tree logs that are deposited on till plain or outwash plain, can help identify the maximum age of moraine formation. The drawback of this method is the large uncertainties associated

Figure 2. Field photos displaying the fresh-looking marginal moraines. (a), (b) and (c) show locations of the Glacier No.1, the Bogeda range, and the Karlik range, respectively, as indicated in **Figure 1(c)**; (d) is located in the Haxilegen site, as marked in **Figure 1(b)**.

with the age [41–43]. In the case of the Tian Shan, lack of organic materials within or nearby moraines often prevents the application of radiocarbon dating. Dendrochronology, which uses tree-ring crossdating technique, is another method to date glacial moraines and can often provide precise age control [44–47]. The year of the innermost ring of living trees that grow on the moraine can indicate the minimum age of the moraine. Tree-ring dated outermost ring of glacially killed trees can be used to cross-validate radiocarbon results. But, the challenge is also the availability of the dating materials: trees are rarely present at margins of glaciers in the Tian Shan. Since the 1990s, cosmogenic exposure dating technique has allowed great improvement in dating glacial landforms. It measures the concentration of the targeted isotope that is only produced by the interaction of cosmic rays with minerals in rocks and then calculates the surface exposure time of the rock [48–51]. Nuclide isotope ^{10}Be is the most widely used one in the application of reconstructing glacial chronology [52]. The advancement in techniques has made this method dating young event on timescales of 100 years possible, for examples, in New Zealand [16], Greenland [53], Switzerland [54] and other places including the Tian Shan [55–57].

Several studies focused on the direct dating of the LIA moraines in the Tian Shan are marked on **Figure 1**. As the climatic conditions shifted with cycles of coldness and warmth during the entire LIA, glaciers in the Tian Shan have been found with two or three moraine ridges that could be identified as LIA moraines [34], indicating different periods of glacial stagnations as glaciers respond to climate changes. In the Chinese Tian Shan, the Chinese Academy of Sciences established the Tian Shan Glaciological Station near the Urumqi Glacier No. 1 in 1959, and thereafter, the Glacier No.1 has become one of a few benchmark glaciers in central Asia (**Figures 1** and **2a**). The dating of the LIA moraines at this site has been conducted by different groups of researchers with various dating methods. Using lichenometry, Chen [58] dated three moraine ridges with minimum formation ages of 1538 ± 20 AD, 1777 ± 20 AD and 1871 ± 20 AD, respectively. Using AMS radiocarbon dating of inorganic carbonate coating of glacial boulders on the outermost moraine, Yi et al. [59] obtained two ages of 450 ± 120 yr and 480 ± 120 yr (calibrated in Ref. [17]), which are averaged and converted to 1535 ± 120 AD, consistent with Chen's results. Our group (authors and collaborators as mentioned in Acknowledgments) conducted cosmogenic exposure dating of the boulders from the fresh moraines at the same site and another site to the west, the Haxilegen Pass (**Figures 1b** and **2d**). The ages of seven samples from the Glacier No. 1 site clustered around 430 ± 110 yr, and the outer moraine at the Haxilegen site was dated to 430 ± 40 yr [57]. Similar ages from different methods provide reliable evidence of the LIA glacial extent, suggesting that the glacier retreated from its LIA maximum extent, 700–900 m from glacier front, approximately 430 years ago at sites in the Chinese Tian Shan. In the Kyrgyz Tian Shan, extensive work of dating LIA moraines has been done using lichenometry. Solomina et al. [36] studied the retreat of 293 glaciers across the Kyrgyz Tian Shan, and lichenometric dates of moraines indicated nearly identical maximum glacier advances occurred in the seventeenth to the mid-nineteenth centuries. Using cosmogenic exposure dating, Koppes et al. [56] measured two ^{10}Be ages for one boulder from the outer moraine and one from the inner moraine in the Ala-Archa Valley, the Kyrgyz Front Range. They are recalculated in [17] to 612 ± 111 yr and 284 ± 75 yr which both belong to the LIA period.

The chronology evidence of the LIA maximum extents at the selected study sites better allows us to use morphologic appearance to interpret potential LIA extent in less-studied areas in the Tian Shan. Indeed, more moraine chronological data across different mountain ranges are needed to examine the timing and the extent of the LIA glaciers and to depict the spatial heterogeneity of LIA glacial events in the region.

3. Reasons for variations

3.1. Climatic driving

The retreat or advance of a glacier is attributed to the amount of snow accumulation and ablation and is a result of an integrated response to climate. Temperature, precipitation, or a composite of climate inputs acts as the major driving force on glacial fluctuations: increased snowfalls or decreased temperature could favour a positive mass balance, making a glacier grow; otherwise, glaciers may recede. Although it is well known that glaciers are a sensitive indicator of climate change, the complexity of the glacier-climate relationship is not an easy puzzle to solve because glaciers are a component linked with many other components in the natural systems and the forcings that influence glacier distribution and changes operate at different scales. It can take several decades for a glacier to respond to some change in climate, and this time lag varies non-linearly and due to many other non-climatic factors such as glacier types and topography [60, 61].

Like in many mountain ranges around the world, the Tian Shan glaciers are retreating or disappearing in response to the increasing atmospheric temperature in the past a few decades. Almost all meteorological stations have observed a warming trend since the 1970s in central Asia [23]. The IPCC Fifth Assessment Report AR5 [62] noted that the warming was particularly strong in winter (November to March), with a rate of 2.4°C per 50 years, in the semi-arid area of Asia. The estimate of the increase in mean annual temperature is about 0.1–0.2°C per decade [62]. Such changes in temperature cause less persistent snow accumulation and prolonged melting season for glaciers. This is likely associated with a weakening or spatial shift of the Siberian high pressure further to the east over the continental Asia in the past century [20, 23].

Precipitation changes driven by the zonal and meridional atmospheric circulation patterns do not show spatially coherent trends in central Asia [28, 63]. Instrumental data of recent decades revealed that the mean annual precipitation generally shows an increasing trend in both northern ranges of the Kyrgyz Tian Shan and the Chinese Tian Shan [64–66], but a decreasing trend in the interior range, the central Tian Shan [23]. Aizen et al. [20] argued that a strengthened westerly flow associated with the warm season precipitation (accounts for most of the annual total) could be the explanation, but the spatial distribution of precipitation decreasing from northwest to southeast reflects the effect of mountain blocking. The regions with increased precipitation undergo glacier retreating too, indicating the impact of current warming on glaciers is not compensated by more precipitation or the sensitivity of glaciers is higher to temperature change in central Asia [28, 67].

The variability and long-time change of the climate system in central Asia are closely related to the interconnections with the large-scale oceanic-atmospheric circulations, such as the El Nino/Southern Oscillation (ENSO), the Atlantic Multidecadal Oscillation (AMO) and the North Atlantic Oscillation (NAO). Several studies have reconstructed millennium-long records of these oscillations and found that the La Niña condition, warm phase AMO and positive NAO are correlated with the above average winter or annual precipitation in arid central Asia [6, 68, 69]. Although a lack of quantity and quality of direct observations hampers the confidence in the assessment of past changes, inferred long-term trends of the indices provided possible explanations of the climatic forcing/mechanisms during the LIA. Evidence derived from climate proxy data is another common approach to extend information on the past climate conditions. Through reconstructions using tree rings, e.g. [70–72], ice core, e.g. [18, 73], eolian sediment, e.g. [74] and lake sediment, e.g. [75], the regional climate during the LIA was depicted as cold and wet in arid central Asia. For example, the Dunde ice core record from the Qilian Mountain, ~500 km southeast of the Tian Shan, identified three long cold periods, around mid-fourteenth century, mid-sixteenth to seventeenth century and the nineteenth century during the LIA [73], which correspond well with the series of glacial advances dated in the moraine chronology [57–59]. IPCC AR5 [62] reported prevailed wetter conditions in central Asia during the LIA compared to the Medieval Climate Anomaly and the twentieth century. On a regional scale, the wet and cold conditions are the primary climatic driving for the LIA glacier expansions in the Tian Shan, but on a more local scale, the non-uniform response to similar climate changes implies other influential factors that need to be taken into account.

3.2. Local factors

Many non-climatic factors influence the development of each glacier and determine how each glacier responds to climate change individually and differently in future. For instance, glacier geometry and topography play an important role in the spatial pattern of glacier change variability. Such factors include glacier size, elevation range, hypsometry (areal distribution by elevation), surface orientation (aspect), slope, shape and surface characteristics (e.g. debris-cover). They vary from region to region and from a glacier to another glacier. Understanding the role of the factors that lead to disparate glacier responses at a local scale will improve our knowledge of glacier-climate interaction and help predict future glacier changes.

Previous studies have explored the impact of a single topographic factor or a combination of factors on the glacier behaviour. For example, Pratt-Sitaula et al. [76] applied cosmogenic ^{10}Be dating on moraines at ice extent maxima in five valleys at Annapurna, Nepal, and found that the glacial asynchrony at neighbouring valleys under no spatial differences in climate is attributed to the effect of hypsometric characteristic. They argued that mountain glaciers with source areas at higher altitude are more likely to lead to glacial advance, while glaciers with lower maximum altitudes tend to retreat when facing a uniform climate change from a cooler-drier late glacial to a warmer-wetter early Holocene [76]. Numerous studies utilized satellite remote sensing data to assess the sequential changes of glaciers during past decades and revealed that in all mountain regions where glaciers exist today, glaciers shrunk considerably over the past 150 years, after the end of the LIA, and many small glaciers have disappeared [32]. Glaciers of different sizes are proven to have strong correlations with the rate of glacier changes, and small glaciers are undergoing stronger retreat compared to large-sized glaciers, e.g. [77–80]. The presence of many small glaciers today are formed due to the rapid disintegration of medium-sized glaciers and are likely to disappear if the warming trend of climate continues [80]. Topographic conditions also matter to the sensitivity and the response time of glaciers to an instantaneous change in climate. Large glaciers that have multiple tributaries and low slope gradient take a longer time to adjust their extent, whereas smaller glaciers respond faster with a higher changing rate; thus, small glaciers are more sensitive to climate change. The influence of the aspect factor can be interpreted in terms of solar radiation receipt, as well as the wind effects to areas with moderate relief [77]. According to the analysis using large data sets from the World Glacier Inventory, Evans [81] measured slope aspect of a total of 66,084 glaciers in 51 regions over the world and found a broadly consistent poleward aspect in middle and high latitudes.

A limited number of studies have been conducted in the Tian Shan to examine the relationship between local factors and glacier changes. In northern Kyrgyz Tian Shan, Bolch [78] compared glacier retreats from 1955 to 1999 among several valleys. He mentioned that the heterogeneity of glacier changes is dependent on the size, as well as the climatic regime divided among northern slopes and southern slopes. Our previous study [82] in the central Chinese Tian Shan was the first to examine the importance of topographic factors to the spatial variability of changes in glacier area and equilibrium line altitude (ELA) since the LIA. The results showed that glacier size and mean elevation range are the two key factors and could explain up to 64% of the relative area change, but the ELA change cannot be well explained by the regression of the local factors [82]. Debris cover is another factor that can have either positive or negative impact on surface ablation because it modifies ice melt rates and the heat conduction in the ablation zone [83]. The thickness of the supraglacial debris was considered to be related to the ablation rate [84]. However, how debris cover influences glaciological processes is quite complex and varies case by case [85, 86]. Most glaciers in the central Tian Shan are the debris-covered type, and glaciers in the eastern Tian Shan are mostly clean-ice type.

A glacier advances or retreats as a result of both local conditions, i.e. topography and its geometry and climate conditions. Understanding these internal and external factors on glacier changes is helpful to discuss the sensitivity of glacier response and is important to interpret paleoclimatic conditions and future glacier evolution. With nearly uniform changes in climate at local scales, more research needs to be focusing on comprehensively investigating the role of local factors played in resulting non-uniform glacier responses.

4. A case study: using statistical models to evaluate the importance of local factors on glacier changes in Eastern Tian Shan

In this case study, we aim to investigate the relationship between the local factors and glacier changes since the LIA in the eastern Tian Shan. The study area (42°40′N–44°00′N, 85°40′E–94°50′E) is entirely bounded within Xinjiang, China, and contains several mountain ranges trending west-east (**Figure 1**). Glaciers exist beyond 3000 m a.s.l., where monthly mean temperatures are −16 to −13°C in winter and 3–5°C in summer [87]. We assumed that at the regional scale, the group of glaciers was impacted by climate change at a similar magnitude since the LIA, and the varied behaviour of any individual glacier reflects the influence from non-climatic factors, including the local topography and glacial geometry.

4.1. Data

The glacier change is defined as the ratio of the areal difference between LIA and the modern glaciers to the area of LIA extents. The extent of modern glaciers was obtained from the Second Glacier Inventory of China [33]. The vector shape file of targeted glaciers in the study area contains attributes of glacier area, elevations and many others following the GLIMS standards. The LIA extent of glaciers is based on the manual delineation of the fresh moraines showing similar features with the sites already dated as LIA moraines in the Tian Shan (see Section 2.3 above). Further details about the delineation can be found in Ref. [88]. In total, 640 LIA glacial extents with corresponding 865 modern glaciers are included in the study. Three sub-regions were defined based on their geographical ranges: the Boro-Eren (BE) range, the Bogeda (BG) range and the Karlik (KL) range. The relative area change for each individual glacier is the dependent variable and seven local factors (glacier area, median elevation, slope, aspect, solar radiation, longitude and latitude) are taken into account as independent variables (**Table 1**). The 1-arc second (~30 m) Shuttle Radar Topography Mission (SRTM) DEM was used to calculate topographic factors that are derived based on elevation. The summary statistics of both the dependent and independent variables used in our model are shown in **Table 2**.

Factor	Variable name	Unit	Summary statistics	Data source and method	Note
Glacier area in LIA	Area	km^2	–	Shapefile	The planar size of glaciers. Converted to logarithmic
Median elevation	elev	m	Median	From DEM	The altitude of glaciers
Surface slope	gslope		Mean	From DEM	Slope can range from 0° to 90°
Surface facing	cosa, sina		Direction-al mean	From DEM	Aspect spans clockwise from 0° (due north) to 360° (again due north). Converted to cosine and sine components using Fourier transformation
Longitude	x	–	Median	Longitude of centroid of each glacier polygon	Longitude determines the west-east location
Latitude	y	–	Median	Latitude of centroid of each glacier polygon	Latitude determines the north-south location
Solar radiation	solar	WH/m^2	Mean	From DEM	Solar radiation is the direct energy input to the glacier surface

Table 1. A list of factors considered in the analysis (modified based on **Table 1** in [89]).

Variable	N	Mean	St. Dev.	Min.	Max.
x	640	88.035	2.512	85.781	94.643
y	640	43.373	0.261	42.975	43.875
solar	640	1254.391	174.661	792.918	1909.040
elev	640	3825.692	172.419	3369	4382
gslope	640	26.562	4.643	13.939	43.036
area	640	1.237	1.602	0.086	17.438
darea	640	49.803	18.823	3.288	95.607
cosa	640	0.629	0.468	−1.000	1.000
sina	640	0.016	0.621	−1.000	1.000

Table 2. Descriptive summary statistics of variables.

4.2. Methods

To minimize the possible bias due to the non-Gaussian distribution of the data, a logarithmic transformation was applied to glacier area. To incorporate circular data of aspect into the calculation, the first harmonic of Fourier transformation was used to convert the aspect degrees to the cosine and sine components. Such a transformation on aspect data was suggested in the past statistical analysis of glaciers: the sine term represents the difference along a west-east direction, whereas the cosine term measures north-south differences [81]. The sub-region codes, 'BE', 'BG' and 'KL', were included as a dummy variable in the statistical regression.

Pearson's pairwise correlation was used to examine if each of the independent variables is significantly positively or negatively correlated with the relative glacier area change and with the others. Considering the multicollinearity between some of the independent variables as well as the possible non-linear relationship between our independent variable and the dependent variable, we used the random forest regression model to identify the factors that have the greatest impact on glacier change. Random forest is similar to regression trees in that a hierarchical classification is used to recursively split data into smaller subsets based on dependent variables. The original data set will be partitioned based on the relationships between the dependent and independent variables, forming different splitting conditions (nodes), and the process of partitioning will not stop until a perfect tree is created, or until pre-defined conditions are met for terminating the expansion of the tree. The random forest does not make an assumption regarding the distribution of the data, and as the bootstrapping procedures are used, the predictability of the random forest model is improved meanwhile the possibility of over-fitting is minimized [90]. The random forest model used the difference in mean square error (MSE) for a test sample with and without a certain variable randomly permuted to determine the importance of each variable. The difference is averaged among all the trees generated and is considered a measure of how the predictive ability is reduced when a certain variable is excluded from the model. Traditional theories concerning how the aforementioned local factors interact with glacier change are often based on scientific abduction, with limited

http://dx.doi.org/10.5772/intechopen.70044

field/lab experiment supporting them, as these types of experiments are labour-intensive, expensive, and time-consuming. We used the randomForest package [90] along with other packages in R Language (http://www.r-project.org) to do all statistical analyses in this study.

4.3. Results and discussion

Among nine variables (**Table 2**), the glacier area, the longitude ('x'), the latitude ('y'), the sine aspect ('sina') and cosine aspect ('cosa') are not normally distributed (**Figure 3**). The longitude variable reflects the distribution of glaciers in the study area, and three clusters can be identified. The cosine component of glacier aspect shows a highly skewed distribution, as a lot of glaciers are north-facing in our study area (cosine ~ 1), whereas the sine component shows a rather uniform distribution, suggesting that glaciers in our data set do not show any clustering pattern concerning east-west facings.

Pearson's pairwise correlation shows that the glacier change is highly correlated with glacier area, elevation, longitude, solar radiation, slope and aspect at 0.01 significance level (**Table 3**). For example, the correlation coefficient between the cosine aspect and the solar radiation is as low as −0.75 ($p < 0.0001$), indicating a high collinearity between the variables. For traditional regression models, the highly correlated variables might result in a potential redundancy using the two variables. The receipt of solar radiation already reflects the aspect factor of a glacier: in the northern hemisphere, a north-facing glacier, which has a higher value of cosine aspect, receives less solar energy compared to a south-facing glacier. If both solar radiation and aspect factors are taken into account, the collinearity between the variables might result in biased estimation of the coefficients in traditional regression models.

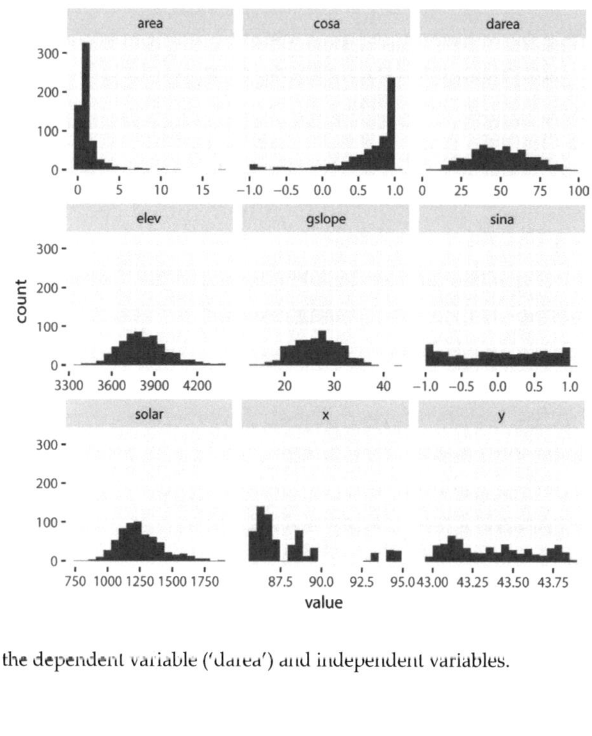

Figure 3. Histograms of the dependent variable ('darea') and independent variables.

	darea	area	elev	x	y	solar	gslope	cosa
area	−0.45***							
elev	−0.53***	0.20***						
x	−0.19***	0.15***	0.22***					
y	−0.03	0.06	−0.19***	−0.13**				
solar	−0.45***	0.23***	0.71***	0.29***	0.01			
gslope	0.38***	−0.36***	−0.05	−0.38***	−0.04	−0.51***		
cosa	0.16***	−0.05	−0.60***	−0.04	−0.05	−0.75***	0.00	
sina	0.11**	−0.04	−0.22***	0.01	0.01	−0.10*	−0.16***	0.14***

***$p < 0.0001$. **$p < 0.01$. *$p < 0.05$.

Table 3. Pearson's correlation matrix.

The random forest regression model calculates an R^2 of 0.541, which suggests that in our study area, 54.1% of the variance in glacier changes can be explained by these local factors. We used 500 trees, the default value for the random forest regression, and the model stabilized when the number of trees exceeds 200 (**Figure 4**). We used a linear regression between the predicted glacier changes calculated from the random forest model and the observed glacier changes to assess the model's predicative ability, and the linear regression had an R^2 of 0.54 ($p < 0.0001$; **Figure 4**). There was no apparent sign of the model over- or under-predicting the glacier change in our study area, and the standard error for our model stayed stationary for different observations.

The relative importance of each independent variable in determining the glacier change is shown in **Figure 4**. The importance of each independent variable in our model is determined by testing how the accuracy of the model will be affected if any individual variable is randomly permuted. The increase in mean squared error (MSE) is calculated for any individual variable when the variable is randomly permuted in the model. The importance of each variable in each tree is calculated, averaged and then normalized using the standard error. Our model suggests that the most important variables affecting glacier changes are elevation and area, which increase the MSE by 60.5 and 52.4%, respectively. Slope, latitude, longitude and region show moderate importance—increasing the MSE by 24.4, 22.6, 22.4 and 19.4%, respectively. Solar radiation is the least important variable controlling glacier change—only increasing the MSE by 19.4%. Elevation and glacier size are two most important factors to control glacier area changes in the random forest model. This result is in good agreement with our previous results built on other statistical models, i.e. multivariate stepwise regression [82] and partial least squares regression [89]. It is commonly recognized that glaciers situated at higher elevations are exposed to colder temperature, thus they tend to recede less; small-sized glaciers tend to be more sensitive to the similar level of climate change compared to larger ones, especially as they expose more proportion of its area under the ELA. Our model ranked the slope as the third most important factor controlling glacier change, followed by latitude, longitude, regional variation, and the solar radiation.

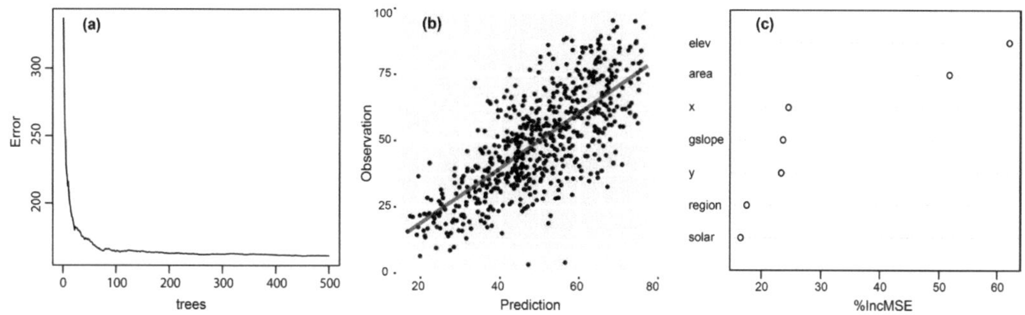

Figure 4. (a) Change of error with increased trees, (b) predicted glacier change and (c) variable importance plot in the random forest model.

The random forest model also produces the representation of the hierarchical classifications which result in high or low glacier changes in our study area. The final regression tree created from the random forest predictions is shown in **Figure 5**. Based on the result, the indication of the mechanism driving glacier changes in our study area can be inferred from the split conditions at each node. We only used split conditions that are significant at the 0.05 level in this figure, since the tree can be morphologically complex when every single node is extended. Also, as the number of nodes becomes larger, our capability to interpret each node becomes very limited. **Figure 5** illustrates the split conditions (nodes) in the random forest model. For example, the first split (node 1) occurs at the elevation of 3795 m; for glaciers with elevation either lower or higher than 3795 m, the next split (node 2 and 17) is based on the size of the glacier.

Low values of glacier retreat occur when the elevation is greater than 3795 m and the area of the glacier is larger than 4.039 km^2 (node 35, glacier change = 20.284%). Larger glaciers that are situated at high altitudes tend to be less sensitive to the impact of climate change, as the majority of the mass of such glaciers are experiencing accumulation, and it is likely that only limited area of such glaciers is experiencing more ablation than accumulation. For glaciers situated at an elevation higher than 3795 m with an area between 2.356 km^2 and 4.039 km^2, the glacier change is also relatively small (28.926%). For glaciers situated higher than 3795 m but with a smaller area (≤2.356 km^2), a smaller slope (≤24.962°) helps limit glacier retreat. However, if the elevation of the glacier is between 3795 m and 3971 m, the latitude of glacier can make a difference in glacier area variation (41.604% for the south of 43.7° and 29.22% for the north of 43.7°.)

Extreme high values of glacier retreat are observed in node 4 (79.112% retreat). This node represents glaciers in the KL region with an elevation lower than 3795 m and smaller than 0.884 km^2. It should also be noted that even for glaciers located higher than 3795 m, high glacier retreat (67.816%, node 27) is observed. This node represents the glaciers in KL region with an area smaller than 2.356 km^2, slope greater than 24.96° and elevation lower than 3884 m. The random forest model is also able to pick up at three split conditions (node 3, 13, 26) where the branches with KL region all show greater glacier changes than the other group, although the glaciers in this region show an overall smaller retreat.

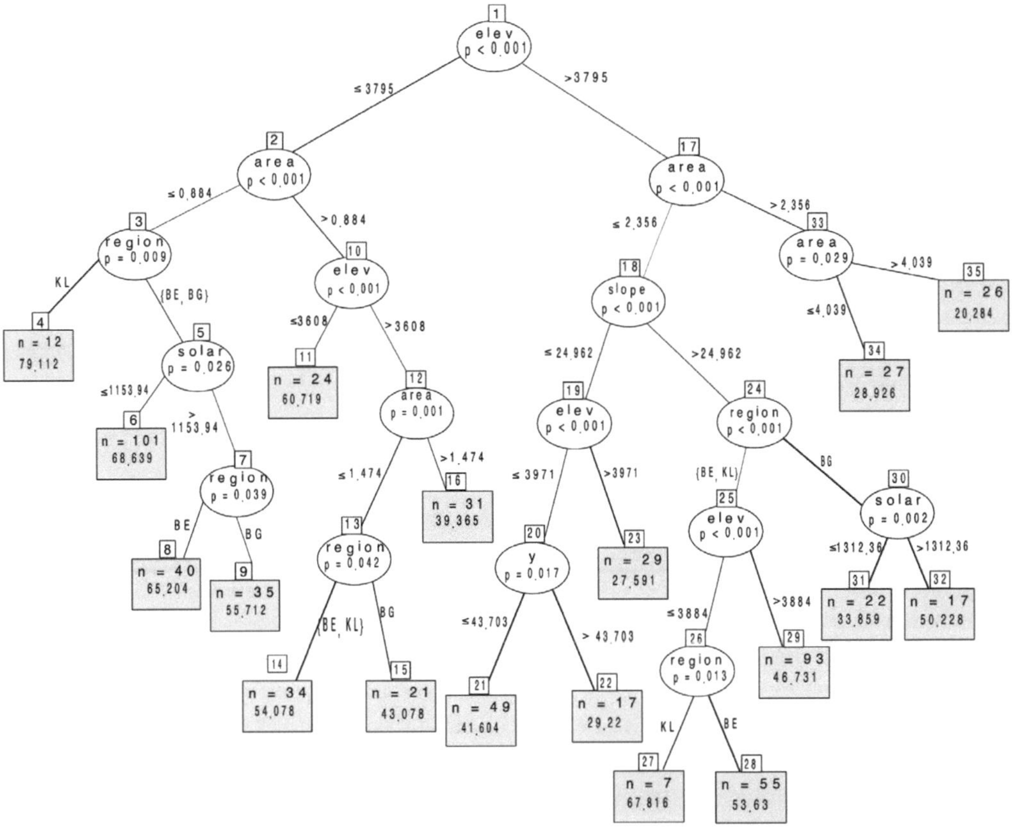

Figure 5. The representative tree of the random forest model. All split conditions are significant at 0.05 level. Grey boxes represent terminal nodes, and the second value is the predicted glacier change at the node.

5. Summary

Located in central Asia, one of the most extreme continental regions on Earth, the Tian Shan, is particularly important because it possesses a high concentration of mountain glaciers which contribute a significant amount of freshwater to populated areas in the lowlands across the bordered countries. These glaciers are also sensitive to climate change and respond to temporal variations in the dominance of major climate systems, such as the mid-latitude westerlies and the Siberian high pressure. The Tian Shan has received little attention compared to other key mountains in the world, partly because of the accessing challenges.

As a WSW-ENE trending ~2500 km arc of mountain system, the Tian Shan includes many ranges across the Kyrgyz Tian Shan in the west and the Chinese Tian Shan in the east. The climatic settings show a high contrast of local differences. On the whole mountain scale, the temperature distribution is quite uniform regardless the change along the elevation, while the precipitation pattern shows strong variations from west to east and from northern slopes to southern slopes. Through some international and national efforts of glacier monitoring, glacier data set/inventories have been collected in the Kyrgyz and Chinese parts using recent remote-sensing data. Overall, there are more than 15,000 glaciers in the whole range, and large glaciers are mostly found in the west, whereas a high number of smaller ones with much less area and volume of ice are found in the east.

Our focus of the Little Ice Age (LIA) allows us to place the contemporary glacier changes under the warming climate into a broader context of glacier history. Although most evidence of the LIA climate and glaciers were assembled in Europe and North America, more and more studies have been conducted in an area like the Tian Shan to reconstruct the past climate conditions and decipher the timing and magnitude of the late Pleistocene glaciations. To date, a few studies have successfully constrained the chronology of the maximum or the successional LIA glacial advances in the Tian Shan, and the dating methods include lichenometry, radiocarbon dating and most recently, cosmogenic nuclide exposure dating. LIA glacial advances occurred around 200 years ago, 450 years ago and 650 years ago based on the moraine chronologies at some sites in the Chinese Tian Shan, but the sequence of the moraine/sub-moraines are not necessarily well-preserved at the front of every glacier. Climate reconstructions from the climate-proxy data, such as tree rings, lake sediments and ice core, revealed a cold and wet climate during the LIA in arid central Asia, and the timing generally agreed with the advances of glaciers. It is well recognized that glaciers respond to changes in climatic conditions, i.e. temperature and precipitation, and during the LIA, the cold, wet climate served as the major driving force to glacier expansions. Admittedly, the mechanism of glacier-climate relationship is complex and difficult to untangle, however, even if assuming similar climate changes at local scales, the variability of glacier changes still exists and implies the influence of other non-climate driving factors, for example, the topographic and geometric settings.

In a case study, we found that the selected local factors (elevation, area, slope, aspect, solar radiation and location) can explain more than 50% of glacier changes since the LIA in the eastern Chinese Tian Shan. This high level of explained variance indicates that local geomorphometric setting is important in determining glacier behaviour and its response to climate change. Among considered factors, glacier size and elevation are the two dominant factors, and this result is in agreement with findings in previous studies. The importance of the other factors is much less than that of these two. At a sub-regional scale, a decreasing trend is observed along a west-east gradient in the correlation between longitude and area change. Such a spatial variation of glacier changes reflected by the three sub-regions might not only be attributed to the difference in their local settings, but also to the gradient in climate systems (i.e. mid-latitude westerlies and the Siberian high pressure). Random forest model is helpful for us to better understand the mechanism between glacier retreat and local factors, and based on the split conditions in random forest model, scenarios resulting in greater or lower glacier retreat are identified.

Future studies in the Tian Shan should consider focusing on several aspects: (1) utilizing high-resolution satellite images to document glacier status and recent glacier changes; (2) developing new sites and improvements in dating techniques to create more well-constrained glacial chronologies; (3) filling the spatial gaps in past climate information by adding more site-specific climate reconstructions across the range and (4) modelling the response of glaciers to climate changes with considerations of local geomorphometric factors.

Acknowledgements

We thank Gengnian Liu, Yixin Chen, and Mei Zhang from Peking University, Jon Harbor and Marc Caffee from Purdue University, and Chaolu Yi from Chinese Academy of Sciences for collaborations on projects of using cosmogenic nuclide technique to date LIA events in the Tian Shan. Sincere thanks are also given to Dr. Danilo Godone for the constructive comments and suggestions to improve the chapter.

Author details

Yanan Li[1]*, Xiaoyu Lu[2] and Yingkui Li[2]

*Address all correspondence to: yanan.li@sdstate.edu

1 Department of Geography, South Dakota State University, Brookings, SD, USA

2 Department of Geography, The University of Tennessee, Knoxville, TN, USA

References

[1] Matthes FE. Report of committee on glaciers, April 1939. Eos, Transactions American Geophysical Union. 1939;**20**(4):518-523

[2] Grove JM. Little Ice Ages: Ancient and Modern (2 volumes). London: Routledge; 2004

[3] Fischer H, et al. Little Ice Age clearly recorded in northern Greenland ice cores. Geophysical Research Letters. 1998;**25**(10):1749-1752

[4] Grove AT. The "Little Ice Age" and its geomorphological consequences in Mediterranean Europe, in the iceberg in the mist: Northern research in pursuit of a "Little Ice Age". Ogilvie AJ and Jónsson T, editors. Netherlands: Springer; 2001. pp. 121-136

[5] Lamb HH. Climatic variation and changes in the wind and ocean circulation - little ice age in the Northeast Atlantic. Quaternary Research. 1979;**11**(1):1-20

[6] Mann ME, et al. Global signatures and dynamical origins of the Little Ice Age and Medieval Climate Anomaly. Science. 2009;**326**(5957):1256-1260

[7] Mosley-Thompson E, et al. Little ice age (neoglacial) paleoenvironmental conditions at siple station, Antarctica. Annals of Glaciology. 1990;**14**:199-204

[8] Pfister C. The little ice age: Thermal and wetness indices for central Europe. Journal of Interdisciplinary History. 1980;**10**(4):665-696

[9] Mann ME. Little ice age. Encyclopedia of Global Environmental Change. 2002;**1**:504-509

[10] Grove JM. The century time-scale. Time-Scales and Environmental Change. New York: Routledge; 1996. pp. 39-87

[11] Kreutz K, et al. Bipolar changes in atmospheric circulation during the Little Ice Age. Science. 1997;**277**(5330):1294-1296

[12] Benedict JB. Recent glacial history of an alpine area in the Colorado Front Range, USA. II: Dating the glacial deposits. Journal of Glaciology. 1968;**7**(49):77-87

[13] Luthi MP. Little Ice Age climate reconstruction from ensemble reanalysis of Alpine glacier fluctuations. Cryosphere. 2014;**8**(2):639-650

[14] Porter SC and Denton GH. Chronology of neoglaciation in the North American Cordillera. American Journal of Science. 1967;**265**(3):177-210

[15] Rabatel A, et al. A chronology of the Little Ice Age in the tropical Andes of Bolivia (16 S) and its implications for climate reconstruction. Quaternary Research. 2008;**70**(2):198-212

[16] Schaefer JM, et al. High-frequency holocene glacier fluctuations in New Zealand differ from the Northern Signature. Science. 2009;**324**(5927):622-625

[17] Xu X and Yi C. Little Ice Age on the Tibetan Plateau and its bordering mountains: Evidence from moraine chronologies. Global and Planetary Change. 2014;**116**:41-53

[18] Yao T, Shi Y, and Thompson L. High resolution record of paleoclimate since the Little Ice Age from the Tibetan ice cores. Quaternary International. 1997;**37**:19-23

[19] Aizen VB, et al. Climatic and hydrologic changes in the Tien Shan, Central Asia. Journal of Climate. 1997;**10**(6):1393-1404

[20] Aizen EM, et al. Precipitation and atmospheric circulation patterns at mid-latitudes of Asia. International Journal of Climatology. 2001;**21**(5):535-556

[21] Benn DI and Owen LA. The role of the Indian summer monsoon and the mid-latitude westerlies in Himalayan glaciation: Review and speculative discussion. Journal of the Geological Society. 1998;**155**:353-363

[22] Kreutz KJ, et al. Oxygen isotopic and soluble ionic composition of a shallow firn core, Inilchek glacier, central Tien Shan. Journal of Glaciology. 2001;**47**(159):548-554

[23] Sorg A, et al. Climate change impacts on glaciers and runoff in Tien Shan (Central Asia). Nature Climate Change. 2012;**2**(10):725-731

[24] Parry ML. Co-authors 2007: Technical Summary. Climate Change 2007: Impacts, Adaptation and Vulnerability. Contribution of Working Group II to the Fourth Assessment Report of the Intergovernmental Panel on Climate Change. Parry ML, Canziani OF, Palutikof JP, van der Linden PJ and Hanson CE, editors. Cambridge, UK: Cambridge University Press; 2006: pp. 23-78

[25] Solomon S, et al. The Physical Science Basis. Contribution of Working Group I to the Fourth Assessment Report of the Intergovernmental Panel on Climate Change. Cambridge, UK and New York, USA: Cambridge University Press; 2007. pp. 235-337

[26] Varis O. Curb vast water use in central Asia. Nature. 2014;**514**(7520):27-29

[27] Aizen VB, et al. Glacier changes in the Tien Shan as determined from topographic and remotely sensed data. Global and Planetary Change. 2007;**56**(3):328-340

[28] Kutuzov S and Shahgedanova M. Glacier retreat and climatic variability in the eastern Terskey–Alatoo, inner Tien Shan between the middle of the 19th century and beginning of the 21st century. Global and Planetary Change. 2009;**69**(1):59-70

[29] Haeberli W, Cihlar J, and Barry RG. Glacier monitoring within the global climate observing system. Annals of Glaciology. 2000;**31**(1):241-246

[30] Bishop MP, et al. Global Land Ice Measurements from Space (GLIMS): Remote sensing and GIS investigations of the Earth's cryosphere. Geocarto International. 2004;**19**(2):57-84

[31] Aizen V, et al. Assessment of glacial area and volume change in Tien Shan (Central Asia) during the last 150 years using geodetic, aerial photo, ASTER and SRTM data. Annals of Glaciology. 2006;**43**:202-213

[32] Shi Y. Concise glacier inventory of China. Shanghai, China: Shanghai Popular Science Press; 2008

[33] Guo W, et al. The Second Glacier Inventory Dataset of China (Version 1.0). Lanzhou, China: Cold and Arid Regions Science Data Center at Lanzhou; 2014

[34] Liu SY, et al. Glacier changes since the Little Ice Age maximum in the western Qilian Shan, northwest China, and consequences of glacier runoff for water supply. Journal of Glaciology. 2003;**49**(164):117-124

[35] Shi Y and Ren J. Glacier recession and lake shrinkage indicating a climatic warming and drying trend in central Asia. Annals of Glaciology. 1990;**14**:261-265

[36] Solomina O, Barry R, and Bodnya M, The retreat of Tien Shan glaciers (Kyrgyzstan) since the Little Ice Age estimated from aerial photographs, lichenometric and historical data. Geografiska Annaler Series A: Physical Geography. 2004;**86**(2):205-215

[37] Beschel RE. Lichens as a measure of the age of recent moraines. Arctic and Alpine Research. 1973:303-309

[38] Loso MG, Doak DF, and Anderson RS. Lichenometric dating of Little Ice Age glacier moraines using explicit demographic models of lichen colonization, growth, and survival. Geografiska Annaler: Series A, Physical Geography. 2014;**96**(1):21-41

[39] Matthews JA and Trenbirth HE. Growth rate of a very large crustose lichen (Rhizocarpon subgenus) and its implications for lichenometry. Geografiska Annaler: Series A, Physical Geography. 2011;**93**(1):27-39

[40] Noller JS and Locke WW. Lichenometry. Quaternary Geochronology: Methods and Applications. The American Geographical Union: Wiley Online Library; 2000. pp. 261-272

[41] Bradley RS and Jonest PD. 'Little Ice Age'summer temperature variations: their nature and relevance to recent global warming trends. The Holocene. 1993;**3**(4):367-376

[42] Hua Q, et al. Radiocarbon in tropical tree rings during the Little Ice Age. Nuclear Instruments and Methods in Physics Research Section B: Beam Interactions with Materials and Atoms. 2004;**223**:489-494

[43] Matthews JA. The late Neoglacial ('Little Ice Age') glacier maximum in southern Norway : New ^{14}C-dating evidence and climatic implications. The Holocene. 1991;**1**(3):219-233

[44] Barclay DJ, Wiles GC, and Calkin PE. Tree-ring crossdates for a First Millennium AD advance of Tebenkof Glacier, southern Alaska. Quaternary Research. 2009;**71**(1):22-26

[45] Bräuning A. Tree-ring evidence of 'Little Ice Age' glacier advances in southern Tibet. The Holocene. 2006;**16**(3):369-380

[46] Wiles GC, Barclay DJ, and Calkin PE. Tree-ring-dated 'Little Ice Age' histories of maritime glaciers from western Prince William Sound, Alaska. The Holocene. 1999;**9**(2):163-173

[47] Wiles GC, Calkin PE, and Jacoby GC. Tree-ring analysis and quaternary geology: Principles and recent applications. Geomorphology. 1996;**16**(3):259-272

[48] Gosse JC and Phillips FM. Terrestrial in situ cosmogenic nuclides: Theory and application. Quaternary Science Reviews. 2001;**20**(14):1475-1560

[49] Ivy-Ochs S and Kober F. Surface exposure dating with cosmogenic nuclides. Eiszeitalter und Gegenwart. 2008;**57**(1/2):179-209

[50] Lal D. Cosmic ray labeling of erosion surfaces: In situ nuclide production rates and erosion models. Earth and Planetary Science Letters. 1991;**104**(2):424-439

[51] Li Y and Harbor J. Cosmogenic nuclides and geomorphology: Theory, limitations, and applications. In: Geomorphology and Plate Tectonics. Hauppauge, NY: Nova Science Publishers, Inc.; 2009. pp. 1-33

[52] Balco G. Contributions and unrealized potential contributions of cosmogenic-nuclide exposure dating to glacier chronology, 1990-2010. Quaternary Science Reviews. 2011;**30**(1-2):3-27

[53] Kelly MA, et al. A ^{10}Be chronology of lateglacial and Holocene mountain glaciation in the Scoresby Sund region, east Greenland: Implications for seasonality during lateglacial time. Quaternary Science Reviews. 2008;**27**(25-26):2273-2282

[54] Schimmelpfennig I, et al. A chronology of Holocene and Little Ice Age glacier culminations of the Steingletscher, Central Alps, Switzerland, based on high-sensitivity beryllium-10 moraine dating. Earth and Planetary Science Letters. 2014;**393**:220-230

[55] Chen Y, et al. Late Quaternary glacial history of the Karlik Range, easternmost Tian Shan, derived from ^{10}Be surface exposure and optically stimulated luminescence datings. Quaternary Science Reviews. 2015;**115**:17-27

[56] Koppes M, et al. Late Quaternary glaciation in the Kyrgyz Tien Shan. Quaternary Science Reviews. 2008;**27**(7/8):846-866

[57] Li Y, et al. Cosmogenic 10 Be constraints on Little Ice Age glacial advances in the eastern Tian Shan, China. Quaternary Science Reviews. 2016;**138**:105-118

[58] Chen J. Preliminary researches on lichenometric chronology of Holocene glacial fluctuations and on other topics in the headwater of Urumqi River, Tianshan Mountains. Science in China Series B-Chemistry. 1989;**32**(12):1487-1500

[59] Yi C, et al. AMS radiocarbon dating of late Quaternary glacial landforms, source of the Urumqi River, Tien Shan—a pilot study of ^{14}C dating on inorganic carbon. Quaternary International. 2004;**121**(1):99-107

[60] Matthews JA and Briffa KR. The 'Little Ice Age': Re-evaluation of an evolving concept. Geografiska Annaler: Series A, Physical Geography. 2005;**87**(1):17-36

[61] Solomina ON, et al. Glacier fluctuations during the past 2000 years. Quaternary Science Reviews. 2016;**149**:61-90

[62] IPCC, Climate Change 2013: The Physical Science Basis, in Contribution of Working Group I to the Fifth Assessment Report of the Intergovernmental Panel on Climate Change, Stocker TF, Qin D, Plattner GK, Tignor M, Allen SK, Boschung J, Nauels A, Xia Y, Bex V, Midgley PM, editors. Cambridge, United Kingdom and New York, NY, USA; 2013. p. 1535

[63] Seneviratne SI, et al. Changes in climate extremes and their impacts on the natural physical environment. In: Managing the Risks of Extreme Events and Disasters to Advance Climate Change Adaptation. Special Report of the Intergovernmental Panel on Climate Change. Cambridge, UK and New York, USA: Cambridge University Press; 2012. pp. 109-230

[64] Chen J, et al. Hydroclimatic changes in China and surroundings during the Medieval Climate Anomaly and Little Ice Age: Spatial patterns and possible mechanisms. Quaternary Science Reviews. 2015;**107**:98-111

[65] Huang W, et al. Interannual precipitation variations in the mid-latitude Asia and their association with large-scale atmospheric circulation. Chinese Science Bulletin. 2013;**58**(32):3962-3968

[66] Shi Y, et al. Recent and future climate change in northwest China. Climatic Change. 2007;**80**(3):379-393

[67] Oerlemans J. Extracting a climate signal from 169 glacier records. Science. 2005;**308**(5722): 675-677

[68] Conroy JL, et al. Unprecedented recent warming of surface temperatures in the eastern tropical Pacific Ocean. Nature Geoscience. 2009;**2**(1):46-50

[69] Trouet V, et al. Persistent positive North Atlantic Oscillation mode dominated the medieval climate anomaly. Science. 2009;**324**(5923):78-80

[70] Cook ER, et al. Tree-ring reconstructed summer temperature anomalies for temperate East Asia since 800 CE. Climate Dynamics. 2013;**41**(11-12):2957-2972

[71] Esper J, et al. Temperature-sensitive Tien Shan tree ring chronologies show multi-centennial growth trends. Climate Dynamics. 2003;**21**(7-8):699-706

[72] Li J, et al. Tree-ring based drought reconstruction for the central Tien Shan area in northwest China. Geophysical Research Letters. 2006;**33**(7):L07715

[73] Thompson LG, et al. Ice core evidence for asynchronous glaciation on the Tibetan Plateau. Quaternary International. 2006;**154**:3-10

[74] Liu WG, et al. Wet climate during the 'Little Ice Age' in the arid Tarim Basin, northwestern China. Holocene. 2011;**21**(3):409-416

[75] Chen FH, et al. Humid Little Ice Age in and central Asia documented by Bosten Lake, Xinjiang, China. Science in China Series D-Earth Sciences. 2006;**49**(12):1280-1290

[76] Pratt-Sitaula B, et al. Topographic control of asynchronous glacial advances: A case study from Annapurna, Nepal. Geophysical Research Letters. 2011; **38**: L24502

[77] Bhambri R, et al. Glacier changes in the Garhwal Himalaya, India, from 1968 to 2006 based on remote sensing. Journal of Glaciology. 2011;**57**(203):543-556

[78] Bolch T. Climate change and glacier retreat in northern Tien Shan (Kazakhstan/Kyrgyzstan) using remote sensing data. Global and Planetary Change. 2007;**56**(1):1-12

[79] Li K, et al. On the relationship between local topography and small glacier change under climatic warming on Mt. Bogda, eastern Tian Shan, China. Journal of Earth Science. 2011;**22**(4):515-527

[80] Paul F, et al. Rapid disintegration of Alpine glaciers observed with satellite data. Geophysical Research Letters. 2004;**31**(21): L21402

[81] Evans IS. Local aspect asymmetry of mountain glaciation: A global survey of consistency of favoured directions for glacier numbers and altitudes. Geomorphology. 2006;**73**(1-2):166-184

[82] Li YA and Li YK. Topographic and geometric controls on glacier changes in the central Tien Shan, China, since the Little Ice Age. Annals of Glaciology. 2014;**55**(66):177-186

[83] Shukla A, Qadir J. Differential response of glaciers with varying debris cover extent: Evidence from changing glacier parameters. International Journal of Remote Sensing. 2016;**37**(11):pp. 2453-2479

[84] Zhang Y, et al. Distribution of debris thickness and its effect on ice melt at Hailuogou glacier, southeastern Tibetan Plateau, using in situ surveys and ASTER imagery. Journal of Glaciology. 2011;**57**(206):pp. 1147-1157

[85] Kayastha RB, et al. Practical prediction of ice melting beneath various thickness of debris cover on Khumbu Glacier, Nepal, using a positive degree-day factor. In: Nakawo M, Raymond CF, and Fountain A, editors. Debris-Covered Glaciers. International Association of Hydrological Sciences: IAHS Press, Wallingford; 2000. pp. 71-81

[86] Scherler D, Bookhagen B, and Strecker MR. Spatially variable response of Himalayan glaciers to climate change affected by debris cover. Nature Geoscience. 2011;**4**(3):156-159

[87] Ye B, et al. The Urumqi River source Glacier No. 1, Tianshan, China: Changes over the past 45 years. Geophysical Research Letters. 2005;**32**(21):L21504

[88] Li Y, et al. Presumed Little Ice Age glacial extent in the eastern Tian Shan, China. Journal of Maps. 2016;**12**(Supp 1):71-78

[89] Li Y, et al. Geomorphometric controls on mountain glacier changes since the Little Ice Age in the Eastern Tien Shan, Central Asia. Annals of the American Association of Geographers. 2017;**107**(2):284-298

[90] Liaw A and Wiener M. Classification and regression by randomForest. R News. 2002;**2**(3):18-22

Chapter 3

The Amazon Glaciers

Rafael da Rocha Ribeiro,
Jefferson Cardia Simões and Edson Ramirez

Additional information is available at the end of the chapter

http://dx.doi.org/10.5772/intechopen.70490

Abstract

In this chapter, we will examine the relationship between the Andean tropical glaciers and the Amazon rainforest, presenting a comprehensive overview on those ice masses that are the headwaters of the Amazon River and examining changes in environmental processes that may affect their mass balance and how they may feedback into the Amazon lowlands environmental processes. The first part of this chapter describes the present glaciological knowledge on these Andean ice masses that flow towards the Amazon drainage basin, about 1666 km^2 (of which 68% are in Peru, 24% in Bolivia and the remaining 8% in Ecuador). The mass balance of these glaciers is strongly dependent on the Amazon hydrological cycle, as water coming from the Atlantic Ocean and recycled though the rainforest is the main source of their precipitation. A second part of the chapter explores how two environmental systems are interconnected and interacted. The third part of chapter examines the present (last 50 years) human-made changes in the Amazon basin and how they may affect the Andean ice masses. These glaciers also hold the best proxy for the Amazon Holocene changes, the record left in the snow and ice chemistry. So, as a complement to this chapter, we review the information on the paleoenvironmental changes found in ice cores in Bolivia and Peru and what they may point about the future of the Andean tropical glaciers.

Keywords: Amazon basin, tropical glaciers, South America

1. Introduction

This chapter examines the relationship between the Andean tropical glaciers and the Amazon rainforest, presenting a comprehensive overview on those ice masses that are the headwaters of the Amazon River and examining changes in environmental processes that may affect their mass balance and how they may feedback into the Amazon lowlands environmental processes.

The first part of this chapter describes the present glaciological knowledge on these Andean ice masses that flow towards the Amazon drainage basin, about 1666 km^2 (of which 68% are in Peru, 24% in Bolivia and the remaining 8% in Ecuador). The mass balance of these glaciers is strongly dependent on the Amazon hydrological cycle because the main source of their snow precipitation are air masses bringing water from the Atlantic Ocean, this water is recycled through the rainforest several times. So, this part of the text also discusses the present atmospheric circulation and how it controls precipitation over the eastern tropical Andean mountains (characterised by a wet and a dry season in Bolivia and Peru) and how signals of changes in the Amazon atmosphere (e.g., pollutants such as black carbon due to biomass burning and trace elements [1]) may be transported to these glaciers. Another important point to consider is how the El Niño-Southern Oscillation (ENSO) phenomenon controls the yearly precipitation variability on these glacier sites.

A second part of the chapter explores how two environmental systems are interconnected and interacted. Not only the existent ice masses are strongly controlled by environmental processes in the rainforest, but the glaciers self-affect their lowlands as providers of sediments [2]. Here, we consider these glaciers as sources of sediments (the Andes tributaries contribute to 90–95% of the Amazon River load [3]), organic matter and nutrients to Amazon basin and how they affect the biochemical, ecological and geomorphological processes. An important point examined is how the melt water variability affects the drainage in the headwaters of the basin (in Bolivia and Peru), and what they represent as water storage and hydric resources for the mountain communities.

The third part of chapter examines the present (last 50 years) human-made changes in the Amazon basin and how they affect the Andean ice masses. The Amazon environment has undergone major changes due to immigration (and the consequent increase of the population in major cities), and intense deforestation (mainly for soy and cattle-culture expansions), affecting the characteristics of the lower atmosphere, changes in land use and land cover also alter and transform the dynamics and formation of clouds [4]. These processes may decrease the precipitation over the Eastern Andes [5], reducing further ice cover already under fast retreat due to climatic warming [6]. One of the main points here to consider is how the loss of mass of these glaciers will affect the water resources of Bolivia and Peru, the former one already under strong hydric stress.

These glaciers also hold the best proxy for the Amazon Holocene changes, the record left in the snow and ice chemistry. So, as a complement to this chapter, we review the information on the paleoenvironmental changes found in ice cores in Bolivia and Peru [1, 7] and what they may point about the future of the Andean tropical glaciers.

2. The present glaciological knowledge on the Andean ice masses that flow towards the Amazon drainage basin

The tropics can be defined as a region where the atmospheric circulation dynamics and the energy conditions present high thermal homogeneity (**Figure 1**). For this reason, the annual

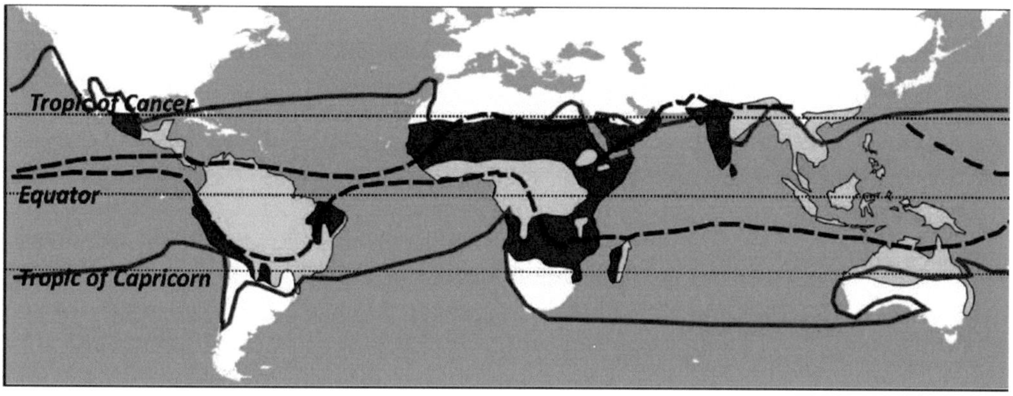

Figure 1. The tropics: in light gray, are the areas that present high precipitation throughout the year; and in dark grey, are the areas with a wet and dry season during the year. The dashed lines identify the seasonal oscillation of the intertropical convergence zone and the continuous lines delimit the tropical zone from the thermal point of view, adapted from Kaser and Osmaston [9].

thermal amplitude is lower than the diurnal temperature variation [8]. In the tropics, unlike the temperate regions, the linear behaviour of the temperature causes the 0.1°C atmospheric isotherm to remain practically at the same altitude, allowing the occurrence of ablation in the glacier terminus throughout the year. At these latitudes, all ice masses are at the pressure melting point and they are classified from the thermal point of view as warm glaciers [9].

In the tropical regions, the conditions of humidity and precipitation (responsible for accumulation) are directly related to the oscillation of the Sun position throughout the year. With a delay of a few weeks in relation to this solar variation, the intertropical convergence zone (ITCZ) position reaches once a year its maximum latitude in a hemisphere, causing a wet season and a very different dry season between these two points of change [9].

In the tropics, glaciers exist in South America (from Bolivia to Venezuela), Africa (Kilimanjaro, Mount Kenya and Rwenzori) and Oceania (West Papua). The Andes have approximately 99% of the ice masses located in the tropics [9]. Tropical South America has an ice cover of 2500 km^2, in which 70% are in Peru, 20% in Bolivia and 10% in Ecuador, Colombia and Venezuela.

The morphology of these sub-equatorial ice masses caused perplexity to the first European travellers. Unlike Alps glaciers, tropical glaciers do not form extensive tongues that flow down the valley walls; they are small in size, like small ice caps, which only cover the mountain peaks [10]. According to Kaser et al. [11], this morphology of the terminal part of the glaciers is due to the continuous annual ablation, unlike what is found in the middle latitudes. Another difference pointed out by the same authors [11] is the glaciers response time to the climatic conditions. In the tropics, the proportion of glaciers that have their area entirely within the ablation sector (glaciers with less than 1 km^2) is higher than in the middle latitudes. Consequently, tropical ice masses respond faster to climate changes than the middle latitude ones.

The distribution of these tropical glaciers is controlled, fundamentally, by two factors: altitude and precipitation. In the former one, the high mountains 'block' the humidity driven by the air masses, providing conditions for the formation of glaciers. The second one is determinant for the equilibrium line altitude (ELA), because the glaciers will only exist where ELA is below the ridges of the mountains [8].

In the Andean mountains, the precipitation is regulated, mainly, by winds coming from the east, which originates on the Atlantic Ocean and mixes to the air masses of the Amazonian origin. The glaciers are distributed predominantly from 12°N to 23°S, in an area tectonically lifted above 5000 m, which 'intercepts' air masses coming from the Amazon forest. This results in a negative precipitation gradient from NE to SW. For example, in northern Bolivia, the ELA is between 5300 and 5800 m a.s.l. Further to the south in this country, even areas 6000 m a.s.l. can be free of ice due to aridity [12].

Figure 2 shows the Amazon River basin and the glaciers that flow to it. For the year 2015, we determined that the 'Amazonian' glaciers covered 1666 km^2 (we obtained this result using the global land ice measurements from space: (GLIMS) database [13]). Of these ice masses,

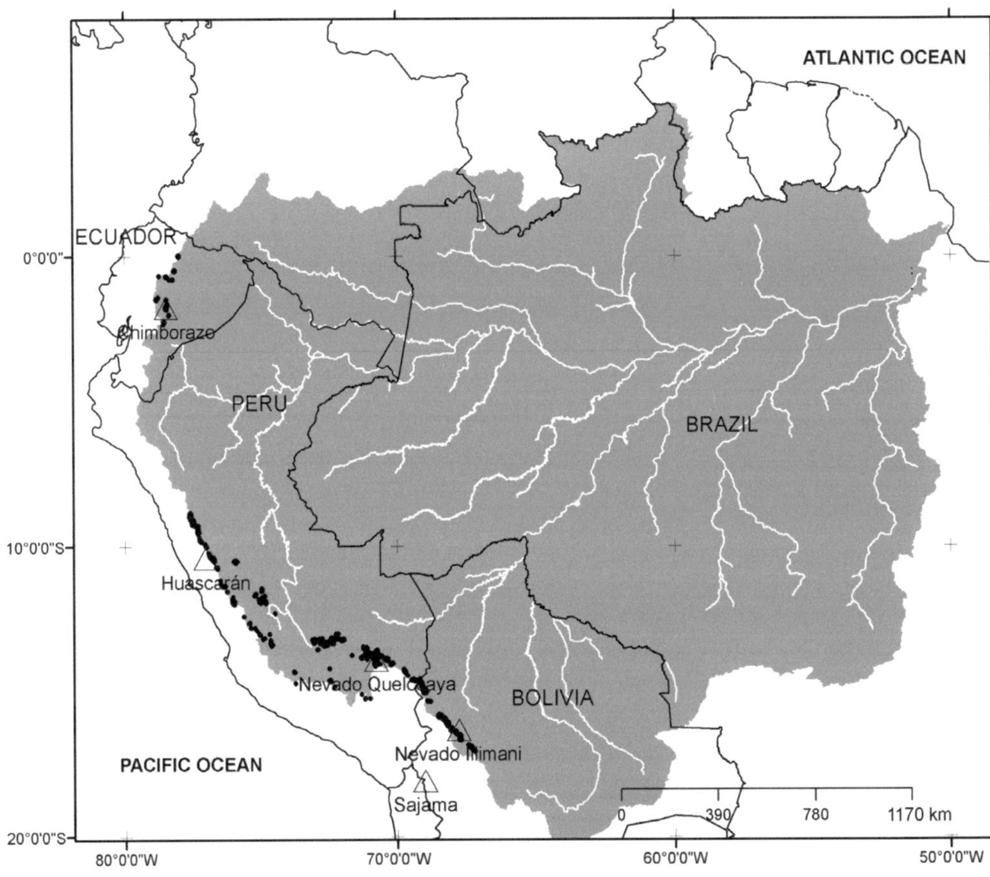

Figure 2. The Amazon glaciers (Andean ice masses flowing towards the Amazon River drainage basin) are shown as black spots. The main ice core sites extracted from the tropical Andes are shown as triangles. The gray area delimits the watershed of the Amazon River.

68% (1129 km²) are in Peru, 24% (397 km²) are in Bolivia and the remaining 8% (139 km²) is in Ecuador. The glacial regime of these ice masses shows yearly humid and dry seasons (Peru and Bolivia) or precipitation throughout the year (Ecuador).

Changes in the Amazonian climate, due to deforestation, indicate that the river sources in the Andes mountains may suffer a significant decrease in their water supply (precipitation) as a result of the reduction of atmospheric humidity [6]. Consequently, the Andean countries may suffer significantly from the rainfall reduction in the Amazon forest [15], as it is already observed, in a similar process, in mountains in western Costa Rica due to deforestation to the east [14]. The Amazonian basin and its forests (by evapotranspiration) affect, therefore, the rainfall in the Andes. This decrease in precipitation, combined with the increase in global temperature, directly influences the behaviour of tropical glaciers [6].

However, the relationship between the hydrological cycle and the climate in South America, especially in the Amazon, still lack further studies. In this region, the weather station networks are very sparse and the lack of high-quality precipitation and river flows data make it difficult to study climate change and climate variability. Thus, it is important to obtain indirect indicators that provide regional environmental information; this point is discussed below when examining the Andean ice cores record.

In the Andean tropical glaciers, the accumulation measured above the 5500 m altitude varies from 0.70 to 1.20 m water equivalent per year. No higher accumulation is observed, which may be related to the low amount of water vapour transported by air masses above 6000 m or to strong winds that do not allow greater accumulations on the summits [16]. On the other hand, larger glaciers can take between 5 and 10 years to respond to changes in the environment. This means that a glacier front movement in a given year depends both on the mass balance in the ablation zone during the same year and on the entire surface of the glacier during previous years. This explains the importance of a long-term analysis on the variations in the glacier front position when studying trends of climate change [10].

The warming rate more than tripled from 1973 to 1998 (from 0.11°C to 0.32–0.34°C/decade); the hottest years were 1997 and 1998, El Niño years [17]. The retraction of several glaciers in the Peruvian Andes [18, 19] was concomitant to this warming. Although it is difficult to pinpoint cause and effect, it seems that climatic warming is the main driver of the rapid retraction and the disappearance of high-altitude ice fields and glaciers in the tropics. These high-altitude tropical ice masses are very close to their melting point [20]; therefore, they respond fast to any change in air temperature. Unlike glaciers in temperate regions in the southern hemisphere, where precipitation on the glacier occurs during the austral winter at low temperatures, in the tropics it happens during the warmer austral summer months. This causes the processes of ablation and accumulation to occur simultaneously, unlike the middle latitude glaciers. This makes your study an excellent indicator of changes in climatic patterns.

The recent glacier retreat in the tropical Andes is the greatest since their maximum extent in the Little Ice Age (LIA, mid-seventeenth to early eighteenth century) [21]. For the past 50 years, the mean mass balance for the Andean glaciers has been more negative at a global scale. This behaviour is attributed, at least partially, to a higher frequency of El Niño events and a warming of the regional troposphere. Several authors project that the smallest glaciers located below 5400 m a.s.l. may disappear before the end of the twenty-first century, given the present climate-warming trend [21, 22].

3. Interconnected and interacted environmental systems

Although greatly reduced in area today when compared to the last glacial maximum (LGM, about 18,000 years before the present), glaciers are important agents of erosion, carving and shaping the high Andean valleys. In addition, in periods when rainfall is scarce, ice melting maintains a minimum water flow and so supplies hydroelectric generation plants, urban centres, etc. Thus, glaciers act as regulators of the hydrological regime in many Andean regions [10].

The tropical Andes includes the headwaters of the Amazon River, which strongly affects its geomorphology, biochemistry and ecology [2]. This can be interpreted in terms of source of sediments, organic matter and nutrients for the lower basin sectors (**Figure 3**) and foreseeable alterations (such as ones due to the construction of dams) can cause major environmental changes [2]. These include the capture of large volumes of sediments and difficulties for the

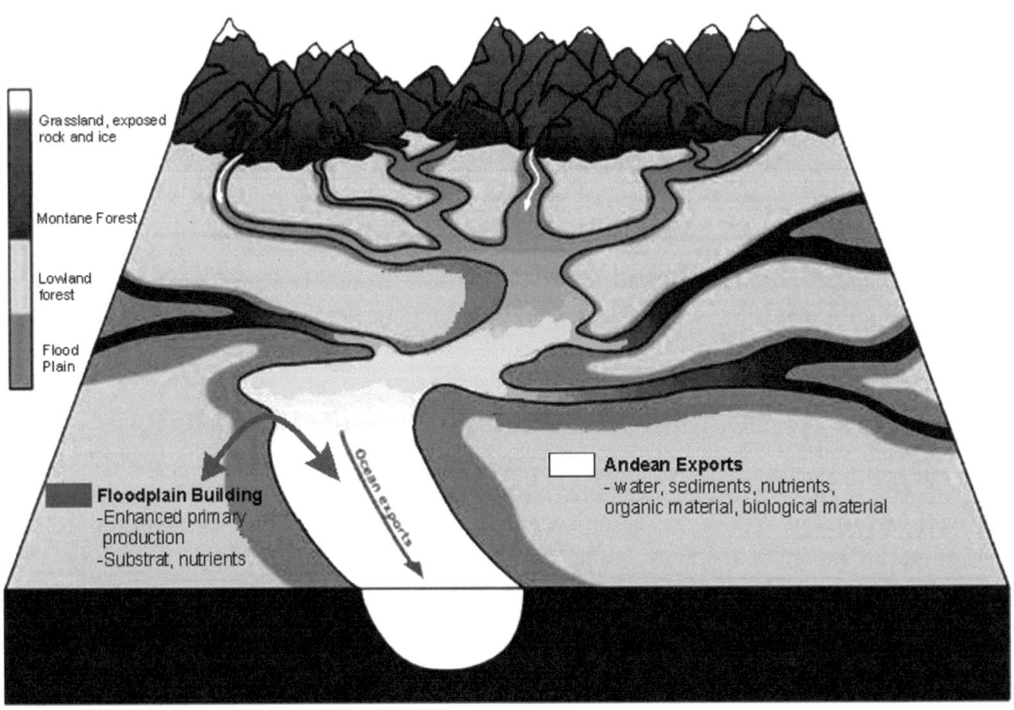

Figure 3. The influence of the Andes on the ecology and biochemistry of the Amazonian forest, adapted from McClain and Naiman [2].

migrating fishes moving from and to the lowlands. It is estimated that less than a quarter of waters of the Amazon basin originate in the Andes, but a greater proportion of the suspended sediments may originate in the mountainous areas. The tributary rivers coming from the Andes provide 90–95% of the suspended sediments of the Amazon River [3]. Taking as an example the Madeira River, which has its headwaters in the high Andes in Bolivia, the total dissolved solids for the Bolivian lowlands at 7 mega grams (Mg) per km^2 per year and the contribution of the Andean basin at 80 Mg per km^2 per year [23]. As for the total sediment load exported from the basin to this river, more than 50% are deposited in the Amazonian plain [24].

There are few regional studies on the sediments flow and the variability of the processes that mobilise them in the Andes. From the transformation of fields to agriculture to the development of hydroelectric plants, this landscape is modified daily. These themes are of great importance, after all, are the most productive components of the Amazon system that are being altered.

The Andean rivers, due to their favourable topography and climate, have a considerable potential for hydroelectric power generation. Practically, half of the regional electricity comes from this power source and small power plants are quite common. Hydroelectricity generation supplies 80, 45, 70 and 40% of the Colombian, Ecuadorian, Peruvian and Bolivian needs, respectively [25]. Studies are developed to identify future demands for energy and to increase the utilisation of the hydroelectric potential in the region with the construction of several small power plants, mainly in Peru and Ecuador [26, 27].

4. Human-made changes in the Amazon basin and how they may affect the Andean ice masses

In the last decades, the Amazonian environment has undergone great changes due to immigration (and the consequent increase in big cities population), and the intense expansion of the soybean cultivation and cattle raising, resulting in an intense deforestation. This is very serious because the forest recycles around 50% of its rainfall, and even an area deforestation of about 30% will not generate sufficient precipitation to maintain itself, generating a negative feedback where 'the greater the forest loss, the less precipitation' [28].

Studies show that many of the South American tropical glaciers have suffered drastic reductions in their areas. To illustrate this, according to the Working Group on Snow and Ice of the International Hydrological Program of the United Nations Educational, Scientific and Cultural Organization (GTHN-PHI-UNESCO), the Cordillera Blanca in Peru have lost 26% of this glacial surface (from 1970 to 2003); for Bolivia, it is estimated a reduction in the order of 50% from the 1970s to the present day; in Ecuador, the glacier inventories points to a 27% area loss from 1997 to 2006 [29].

One of the factors responsible for the disappearance of these ice masses could be the increase in the mean annual temperature in the tropical Andes by approximately +0.8°C from 1970 to

2000 [30]. This is due to the increase in rainfall (in lieu of snow) in the lower sectors of the glaciers. Thus, ice is more exposed (<albedo), increasing the amount of energy absorbed by the glacier surface and intensifying its melting [31]. This situation means that small glaciers (<1 km^2), located at low altitudes, do not recover their deficit, even during the colder years. Therefore, they are in serious danger of disappearing [16].

The Amazonian climate is regular, being exceptionally modified during El Niño events, which changes the rain regime bringing dry periods in the rainy season (December/January/February) [32]. During this climatic event, the intertropical convergence zone (ITCZ) moves farther north than its normal position on the tropical Atlantic Ocean. As a consequence, we have weaker trade winds from the northeast that reduce the moisture that penetrates into the interior of the Amazon region, and so, inhibiting the formation of convective activity. Another consequence is the increase in temperature and carbon dioxide levels. As successive dry years (due to El Niño) reduce the vegetation, primary photosynthesis increases by human intervention, fires in the forests, mainly in Brazil and Bolivia [33].

For precipitation, however, no trend is evident and its variability may be associated with regional particularities, marked by its relation to ENSO phenomenon. El Niño events are associated with temperatures 1–3°C above the average in the Central Andes, which causes an increase in the melting of glaciers and decrease in cloudiness, which end up keeping albedo values low [34]. In hydrographic basins that have essentially a glacial regime, melting flows dominate during El Niños events. On the other hand, in the hydrological basins with few glacierized areas, the water runoff increase by melting is not enough to compensate for the deficits produced by precipitation scarcity (i.e. Bolivia and southern Peru). The greater frequency of El Niño events, since the 1970s, associated with the increase in the mean annual temperature, explains part of the retreat of the tropical Andean glaciers [21, 29].

The irregular behaviour of the El Niño–La Niña events is one of the main uncertainties in the climate projections for the Andes. Thus, a reliable prediction for this region is difficult, since the runoff depends heavily on the occurrence of these events. El Niño accelerates the retreat of the glaciers by rising temperatures (in Bolivia, Peru and Ecuador) or by the decrease of precipitation (in Bolivia and southern Peru). Therefore, the models for this region should be treated with great care [35].

Current projections suggest that the mean temperatures in the Andes can increase by 4.5–5°C in the twenty-first century [36]. As for the disappearance of the glaciers, there will initially be an increase in the flows of the rivers in the basins supplied by melting water, followed by a drastic decrease in the volume and in the regularity of the water resources, and finally, the hydrological regime will become more and more nival-pluvial.

The flow volume of the rivers in the Cordillera Blanca (Peru) drainage basin may disappear between 2175 and 2250 [37]. In Colombia, the retraction of the glaciers will result in water availability problems between 2015 and 2025 [38]. In Ecuador, the reduction of melting water will not only affect headwater areas, but also especially the production of water in the páramos (moorlands) and existing aquifers [31]. The issue becomes even more complex when we take into account the domestic water use, countries like Ecuador and Colombia depend

http://dx.doi.org/10.5772/intechopen.70490

fundamentally on the flow of páramos water (this is a mountain ecosystem, which has great potential for water storage) of the Andes [36, 39, 40]. In addition, in Bolivia and southern Peru, the water source predominantly comes from the high Andean rivers.

As most tropical glaciers are less than 200 m thick and have a small ice volume, their total melting would cause an insignificant increase in sea level (±0.1 mm). If we consider the melting of all mountain glaciers in the world, this increase would be of only 24 cm [16]. If the entire ice sheets of Antarctica and Greenland melted, it would produce an increase of approximately 72 m in the sea level. Thus, the study of tropical glaciers as regulators of regional water availability, as well as 'production' of sediments and nutrients for the Amazon basin, is of great importance, but the impact on the sea level is negligible.

In future scenarios of climate change, it is anticipated that many of the small mountain glaciers located in low-lying areas of the tropics will disappear in few decades. In addition to the decline in water resource, shrinkage of tropical ice masses will also create hazards, such as instability in mountain slopes, glacier detachment, avalanches and lagoon overflow, as well as changes in ecosystems [36, 39, 41]. An increase in the frequency of extreme events (e.g. storms) may have implications for slope stability, with risks for cities in areas at lower elevations. The effects on the nutrient cycle are still uncertain [2].

The expected environmental changes will cause profound modifications in the flow patterns of many rivers. These modifications will reduce the rivers capacity to provide ecosystem services (sources of water and food, recreation, assimilation of the páramos and flow control) [39, 42]. An alternative to this scenario would be the creation of water regulations and policies that recognise the need to maintain specific flow regimes to sustain ecosystems [43].

We highlighted the importance of the South American tropical glaciers for the Amazon basin at a regional scale. Although these glaciers cover a small area (about 2500 km^2), the impact of the environmental changes on them will have consequences for the Amazonian rivers. However, more studies are needed to determine the processes and scales of these modifications and develop mechanisms to protect the ecosystems associated with the Andean glaciers.

5. Andean glaciers also hold the best proxy for the Amazon Holocene changes, the record left in the snow and ice chemistry

Reliable climatic data from the Amazon are still deficient; a detailed monitoring of large areas of the Amazon basin requires a dense network of rain gauges, which in some cases is not feasible by topography and forest [44]. Therefore, any analysis in this region should be taken with caution [45, 46]. On the other hand, ice cores provide archives of the past climate record, on the climatic forcing at the time of its deposition (as changes in solar activity) and volcanic eruptions.

Several ice cores were extracted and analysed in the Andes for information on the environmental conditions of the tropics (**Figure 2**): in southern Peru (Quelccaya ice cap, 13°56′ S,

70°50′ W, 5670 m a.s.l.); in the Cordillera Branca, Peru (Huascarán mountain, 9°07′ S, 77°37′ W, 6048 m a.s.l.); in western Bolivia (Sajama ice cap, 18°06′ S, 68°53′ W, 6542 m a.s.l.) and in the central sector of the Andes (Nevado Illimani, 16°37′ S, 67°46′ W, 6350 m a.s.l.) [11, 47–51]. The Nevado Illimani record is particularly interesting for this chapter, as it is less than 500 km of the Amazon rainforest (receiving by advection the humid masses from this region), providing information on the composition and evolution of the atmospheric chemistry of the Amazon region [11, 51].

The annual climate over the tropics is dominated by two well-defined seasons (summer/wet and winter/dry) and the glaciers of the central Andes are fed during the wet season by precipitation coming from the Amazon basin. Therefore, we can consider the snow and ice layers of these glaciers as indirect indicators (proxies) of the environmental conditions of the South America.

The snowfall of the Illimani Nevado shows traces of biomass emission (e.g. ammonia, acetate, potassium) as a dominant contribution coming from the Amazon basin [52]. It is important to notice that water vapour recycles several times along its path from the Atlantic through the Amazon basin before precipitating in the Andes glaciers.

An alternative technique for the study of the climatic variables of a region can be based on the ratio in the stable isotope ratios (δD and $\delta^{18}O$) in rainwater and snow [53, 54]. It is known that the present proportion of these elements is controlled by meteorological parameters (temperature, precipitation volume, etc.), which allows reconstructing/estimate the climatic conditions in the past. This technique also allows the identification of the air masses that undergo precipitation [47, 51, 53, 55, 56].

The analysis of the four ice cores extracted from the Andean tropics (Huascarán, Quelccaya, Illimani and Sajama) was based (mostly) on the information deduced from the content of hydrogen and oxygen isotopes. The results showed good consistency among the records, suggesting a similar climatic history for the twentieth century [51]. Initially, Lonnie Thompson from the Ohio State University used the isotopic oxygen content record to analyse temperature changes based on similar records for high latitudes [48, 49]. Notwithstanding, when comparing the meteorological records with the isotopic records for the tropical Andes, other authors [55, 57] identified a strong correlation between changes in precipitation and changes in oxygen values. These authors come to the conclusion that changes in precipitation origin and amount are more important than temperature (as originally proposed by for high latitudes [53]) as controller of the isotopic ratios in the tropics.

A study of the Huascarán mountain ice cores identified that zonal wind variations over South America at 500 hPa are closely related to the interannual variations in the $\delta^{18}O$ values [48]. This suggests that the sea surface temperature (SST) in the western tropical Atlantic influences the circulation at 500 hPa, the moisture isotopic fractionation process along its Amazon pathway and so, the $\delta^{18}O$ precipitation at the ice core site.

It is not known whether the temperature effect or the amount effect predominates in the isotopic signal in tropical glaciers [58]. The authors determined the correlation of $\delta^{18}O$ with the ENSO variability in the Illimani Nevado ice core, and identified that more negative values of this isotope ratio coincide with Pacific SST (sea surface temperature) below average. But this interpretation on stable oxygen isotopic ratios should be taken with caution, though it is assumed that the lowlands to the east of the Andes and the Atlantic Ocean are the sources of moisture for precipitation in the glaciers, the local conditions such as condensation and water recycling should not be overlooked.

The Atlantic Ocean, by the trade winds circulation, 'feeds' moisture into the Andean snow-fields, so the SST variability in the Andes precipitation is first 'filtered' across the Atlantic sector. When the Pacific is hot/cold, surface temperature anomalies are redistributed to the tropical Atlantic basin, so the moisture flow responsible for the precipitation in the Andes can be remotely controlled by tropical Pacific conditions [59].

Briefly, we can say that the primary moisture source for precipitation in the tropical glaciers of the Amazon basin is the Atlantic Ocean. So, the record of an Andean ice cores is a link between the meteorological processes of the Amazon basin, Andes and the Atlantic Ocean. Although the $\delta^{18}O$ variations in the tropical Peru are strongly controlled to ENSO variability [7, 60] on an interannual scale, the Amazon source has remained consistent since the late glacial stage (LGS) [61] and all Peruvian-Bolivian high-altitude ice core records reflect the Amazonian source.

The low nitrate record from the Huascarán and Sajama ice cores point to a less extensive Amazon Basin forest [20] during the LGS. This reduction of forest cover and expansion of savannahs are consistent with greater abundance of eolian particles at the Huascarán core site [20, 48].

For the little ice age (LIA, generally accept as a cool period from 1200 to 1800 C.E.), elevated concentrations of nitrate (NO_3^-) and relative low ^{18}O at the Quelccaya site are observed [7], interpreting these observations as a consequence of more moist condition to the south over the Amazon Basin. Furthermore, they observed that the LIA is a marked feature in three ice core records (Huascarán, Quelccaya and Illimani). During the past 200 years, $\delta^{18}O$ have increased, reaching the highest values for the past 6000 years [7].

Finally, the Illimani ice core records positive trends in trace species of anthropogenic origin (Cu, As, Zn, Cd, Co, Ni and Cr) from the beginning of twentieth century [1], high amounts of biomass emission tracers (ammonium, formate, acetate, oxalate, potassium) are found at the same core [62].

Author details

Rafael da Rocha Ribeiro[1]*, Jefferson Cardia Simões[1] and Edson Ramirez[2]

*Address all correspondence to: r.ribeiro@ufrgs.br

1 Instituto de Geociências, Centro Polar e Climático – CPC, Universidade Federal do Rio Grande do Sul – UFRGS, Porto Alegre, RS, Brazil

2 Instituto de Hidráulica e Hidrologia, Universidad Mayor de San Andrés, La Paz, Bolivia

References

[1] Correia A, Freydier R, Delmas RJ, Simões JC, Taupin JD, Dupré B, Artaxo P. Trace elements in South America aerosol during 20th century inferred from a Nevado Illimani ice core, Eastern Bolivian Andes (6350 m a.S.L.). Atmospheric Chemistry and Physics. 2003;**3**:2143-2177

[2] McClain ME, Naiman RJ. Andean influences on the biogeochemistry and ecology of the Amazon River. Bioscience. 2008;**58**:325-338

[3] Meade RH, Dunne T, Richey JE, Santos UDM, Salati E. Storage and remobilization of sediment in the lower Amazon River of Brazil. Science. 1985;**228**:488-490

[4] Pielke RA. Mesoscale Meteorological Modeling. 2nd ed. San Diego: Academic Press; 2002:676 p

[5] Paiva EMCD, Clarke RT. Time trends in rainfall records in Amazônia. Bulletin of the American Meteorological Society. 1995;**76**:2203-2209

[6] Bunyard P. Climate and the Amazon: Consequences for our planet. Proceedings of the First International Conference on the Impact of Climate Change on High-Mountain Systems, Bogota, 21-23 November 2005, edited by Ceballos, JL, Huggel C, Saldarriaga G, Yepes L, IDEAM, Bogota, 109-121. 2005. 294p.

[7] Thompson LG, Mosley-Thompson E, Davis ME, Zagorodnov VS, Howat IM, Mikhalenko VN, Lin PN. Annually resolved ice core records of tropical climate variability over the past 1800 years. Science. 2013;**340**(6135):945-950

[8] Clapperton CM. Quaternary Geology and Geomorphology of South America. Amsterdam: Elsevier; 1993:779 p

[9] Kaser G, Osmaston O. Tropical Glaciers. Cambridge: Cambridge University Press; 2002: 210 p

[10] Francou B, Pouyaud B. Métodos de observación de glaciares en los Andes tropicales. Mediciones de terreno y procesamiento de datos. Great Ice, IRD: France; 2004:238 p

[11] Kaser G, Ames A, Zamora M. Glacier fluctuations and climate in the cordillera Blanca. Annals of Glaciology. 1990;**14**:136-140

[12] Hastenrath S. Observations on the snowline in the Peruvian Andes. Journal of Glaciology. 1967;**6**:541-550

[13] GLIMS: Global Land Ice Measurements from Space: Monitoring the World's Changing Glaciers. Available from: https://www.glims.org/ [Accessed: July 2017].

[14] Ray DK, Nair US, Lawton RO, Welch RM, Pielke RA Sr. Impact of land use on costa Rican tropical montane cloud forests: Sensitivity of orographic cloud formation to deforestation in the plains. Journal of Geophysical Resource. 2006;**111**. Pages 2 and 15. DOI: 10.1029/2005JD006096

[15] Lean J, Rowntree PR. AGCM simulation of Amazonian deforestation on climate using an improved canopy representation. Quarterly Journal of the Royal Meteorological Society. 1993;**119**:509-530

[16] Andina C. ¿El fin de las cumbres nevadas? Glaciares y Cambio Climático en la Comunidad Andina. Lima: Secretaría General de la Comunidad Andina, Programa de las Naciones Unidas para el Medio Ambiente, Oficina Regional para América Latina y el Caribe y la Agencia Española de Cooperación Internacional; 2007:103 p

[17] Vuille M. Atmospheric circulation over the Bolivian altiplano during dry and wet periods and extreme phases of the southern oscillation. International Journal of Climatology. 1999;**19**:1579-1600

[18] Ames A. A documentation of glacier tongue variations and lake development in the cordillera Blanca, Peru. Zeitschrift fur Gletscherkunde und Glazialgeologie. 1998;**34**:1-36

[19] Thompson LG, Mosley-Thompson E, Davis ME, Lin PN, Henderson KA, Mashiotta TA. Tropical glacier and ice core evidence of climate change on annual to millennial time scales. Climate Change. 2003;**59**:137-155

[20] Thompson LG, Mosley-Thompson E, Henderson KA. Ice-core palaeoclimate records in tropical South America since the last glacial maximum. Science. 2000;**15**:377-394

[21] Rabatel R, Francou B, Soruco A, Gomez J, Cáceres B, Ceballos JL, Basantes R, Vuille M, Sicart JE, Huggel C, Scheel M, Lejeune Y, Arnaud Y, Colle M, Condom T, Consoli G, Favier V, Jomelli V, Galarraga R, Ginot P, Maisincho L, Mendoza J, Meeneegoz M, Ramirez E, Ribstein P, Suzarez W, Villacis M, Wagnon P. Current state of glaciers in the tropical Andes: A multi-century perspective on glacier evolution and climate change. The Cryosphere. 2013;**7**:81-102

[22] Thompson LG, Mosley-Thompson E, Davis ME, Brecher HH. Tropical glaciers, recorders and indicators of climate change, are disappearing globally. Annals of Glaciology. 2011;**52**(59):23-34

[23] Guyot JL, Quintanilla J, Callidonde M, Calle H. Distribución regional de la hidroquímica en la cuenca Amazonica de Bolivia. In: Roche MA, Bourges J, Salas E, Diaz C. Programa Hidrológico y Climatológico de la cuenca Amazonica de Bolívia: Seminário PHICAB. La Paz: ORSTOM, CONAPHI, IHH; 1992. pp. 135-144.

[24] Guyot JL, Bourges J, Hoorelbecke R, Roche MA, Calle H, Cortes J, Barragan MC Exportation de matières en suspension des Andes vers l'Amazonie par le Rio Béni, Bolivie. In: Bordas P, Walling DE, editors, Sediment Budgets. Wallingford: IAHS; 1988.**174**:443-451.

[25] Anderson EP, Encalada AC, Maldonado-Ocampo JA, McClain ME, Ortega H, Wilcox BP. Environmental flows: A concept for addressing effects of river alterations and climate change in the Andes climate change: Evidence and future scenarios for the Andean region. In: Herzog SK, Martínez R, Jófgensen PM, Tiessen H, editors. Climate Change and Biodiversity

in the Tropical Andes. São Paulo: Inter-American Institute for Global Change Research (IAI) and Scientific Committee on Problems of the Environment (SCOPE); 2011:348 p

[26] Pelaez-Samaniego MR, Garcia-Perez M, Cortez LAB, Oscullo J, Olmedo G. Energy sector in Ecuador: Current status. Energy Policy. 2007;**35**:4177-4189

[27] Vergara W, Deeb A, Valencia A, Bradley R, Francou B, Zarzar A, Grünwldt A, Haeussling S. Economic impacts of rapid glacier retreat in the Andes. EOS Transactions, American Geophysical Union. 2007;**88**:261-263

[28] Marengo JA, Nobre CA, Chou SC, Tomasella J, Sampaio G, Alves LM, Obregón GO, Soares WR, Betts R, Kay G. Riscos das Mudanças Climáticas no Brasil: Análise conjunta Brasil-Reino Unido sobre os impactos das Mudanças Climáticas e do Desmatamento na Amazônia. São José dos Campos: INPE. 2011; 55 p

[29] GTHN-PHI-UNESCO – Grupo de Trabajo de Nieves y Hielos del Programa Hidrológico Internacional de United Nations Educational, Scientific and Cultural Organization. Declaración de Manizales sobre "Glaciares y Cambio Climático". In: López Arenas CD, Ramírez Cadena J, org. Glaciares, nieves y hielos de América Latina: Cambio climático y amenazas. Bogotá, Ingeominas. Vol. 1. 2009. pp. 13-16.

[30] Vuille M, Bradley R. Mean temperature trends and their vertical structure in the tropical Andes. Geophysical Research Letters. 2000;**27**(23):3885-3888

[31] Favier V, Coudrain A, Cadier E, Francou B, Ayabaca E, Maisincho L, Pradeiro E, Villacís M, Wagnon P. Evidence of groundwater flow on Antizana icecovered volcano, Ecuador. Hydrological Sciences Journal. 2008;**53**:278-291

[32] Fisch G, Marengo JA, Nobre CA. Uma revisão geral sobre o clima da Amazônia. Acta Amazônica. 1998;**28**(2):101-126

[33] Cochrane MA, Laurence WF. Fire as a large-scale edge effect in Amazonian forests. Journal of Tropical Ecology. 2002;**18**:311-325

[34] Francou B, Pizarro L. El Niño y la sequía en los Altos Andes (Perú y Bolivia). Bulletin de l'Institut Français d'Études Andines. 1985;**14**(1-2):1-18

[35] Marengo JA, Pabéon JD, Diaz A, Rosas G, Ávalos G, Montealegre E, Villacis M, Solman S, Rojas M. Climate Change: Evidence and future scenarios for the Andean region. In: Climate Change and Biodiversity in the Tropical Andes. Herzog SK, Martínez R, Jófgensen PM, Tiessen H. . São Paulo: Inter-American Institute for Global Change Research (IAI) and Scientific Committee on Problems of the Environment (SCOPE); 2011. 348 p

[36] Vuille M, Francou B, Wagnon P, Juen I, Kaser G, Mark BG, Bradley RS. Climate change and tropical Andean glaciers: Past, present and future. Earth Science Reviews. 2008;**89**:79-96

[37] Pouyaud B, Zapata M, Yerren J, Gomez J, Rosas G, Suarez W, Ribstein P. Avenir des ressources en eau glaciaire de la Cordillère Blanche. Hydrological Sciences Journal. 2005;**50**:999-1022

[38] IDEAM. Los glaciares colombianos, expresión del cambio climático global. Bogotá, Instituto de Hidrología, Meteorología y Estudios Ambientales; 2000 19 p

[39] Buytaert W, Celleri R, De Bievre B, Hofstede R, Cisneros F, Wyseure G, Deckers J. Human impact on the hydrology of the Andean páramos. Earth Science Reviews. 2006;**79**:53-72

[40] Viviroli D, Archer D, Buytaert W, Fowler HJ, Greenwood GB, Hamlet AF, Huang Y, Koboltschning G, Litaor MI, Lopez-Moreno JI, Lorentz S, Schaedler B, Schreier H, Schwaiger K, Vuille M, Woods R. Climate change and mountain water resources: Overview and recommendations for research management and policy. Hydrology and Earth System Sciences. 2011;**15**:471-504

[41] Francou B, Vicent C. Les glaciers vont-ils disparaître? Les glaciers face au changement climatique. France: IRD Editions et CNRS Editions; 2007

[42] Bradley RS, Vuille M, Diaz HF, Vergara W. Climate change: Threats to water supplies in the tropical Andes. Science. 2006;**312**:1755-1756

[43] Poff NL, Richter BD, Arthington AH, Bunn SE, Naimam RJ, Kendy E, Acreman M, Apse C, Bledsoe BP, Freeman MC, Henriksen J, Jacobson RB, Kennen JG, Merritt DM, O'KEEFFE JH, Olden JD, Rodgers K, Tharme RE, Warner A. The ecological limits of hydrologic alteration (ELOHA): A new framework for developing regional environmental flow standards. Freshwater Biology. 2010;**55**:147-170

[44] Angelis CF, McGregor GR, Kidd C. Diurnal cycle of rainfall over the Brazilian Amazon. Climate Research. 2004;**26**:139-149

[45] Marengo JA. Interdecadal variability and trends of rainfall across the Amazon basin. Theoretical and Applied Climatology. 2004;**78**:79-96

[46] Satyamurty P, Castro AA, Tota J, Gularte LES, Manz AO. Rainfall trends in the Brazilian Amazon Basin in the past eight decades. Theoretical and Applied Climatology. 2010;**99**:139-148

[47] Thompson LG, Mosley-Thompson E, Bolzan JF, Koci BR. A 1500-year record of tropical precipitation in ice core from the Quelccaya ice cap, Peru. Science. 1985;**229**:971-973

[48] Thompson LG, Mosley-Thompson E, Davis ME, Lin PN, Henderson KA, Cole-Dai J, Bolzan JF, Liu KB. Late glacial stage and Holocene tropical ice core records from Huascarán, Peru. Science. 1995;**269**:46-50

[49] Thompson LG, Davis ME, Mosley-Thompson E, Sowers TA, Henderson KA, Zagorodnov VS, Lin PN, Mikhalenko VN, Campen RK, Bolzan JF, Cole-Dai J, Francou B. A 25,000 year tropical climate history from Bolivian ice cores. Science. 1998;**282**:1858-1864

[50] De Angelis M, Simões JC, Bonnaveira H, Taupin JD, Delmas RJ. Volcanic eruptions recorded in the Illimani ice core (Bolivia): 1918-98 and Tambora periods. Atmospheric Chemistry and Physics. 2003;**3**:1725-1741. DOI: 10.1029/2003JD003JD003623.

[51] Ramirez E, Hoffmann G, Taupin JD, Francou B, Ribstein P, Caillon N, Ferron FA, Landais A, Petir JR, Pouyaud B, Schotterer U, Simões JC, Stievenard MA. New Andean deep

ice core from Nevado Illimani (6350 m), Bolivia. Earth and Planetary Science Letters. 2003;**212**:337-350

[52] Correia A. Histórico da deposição de elementos traço na Bacia Amazônica Ocidental ao longo do século XX. [PhD thesis]. São Paulo, Brazil: Universidade de São Paulo; 2003

[53] Dansgaard W. Stable isotopes in precipitation. Tellus. 1964;**16**:436-468

[54] Bowen GJ, Revenaugh J. Interpolating the isotopic composition of modern meteoric precipitation. Water Resources Research. 2003;**39**(10):1299. DOI: 10.1029/2003WR002086.

[55] Hoffmann G, Ramirez E, Taupin JD, Francou B, Ribstein P, Dela R, Durr H, Gallaire R, Simões JC, Schotterer U, Stievenard M, Werner M. Coherent isotope history of Andean ice cores over the last century. Geophysical Research Letters. 2003;**30**(4):1179-1183

[56] Kutschera W. The role of isotopes in environmental and climate studies. Nuclear Physics. 2005;**A 752**(1):645-648

[57] Hardy DR, Vuille M, Bradley RS. Variability of snow accumulation and isotopic composition on Nevado Sajama, Bolivia. Journal of Geophysical Research. 2003;**108**(D22):4693

[58] Knüsel S, Brütsch S, Henderson KA, Palmer AS, Schwikowski M. ENSO signals of the twentieth century in an ice core from Nevado Illimani, Bolivia. Journal of Geophysical Research. 2005;**110**:D01102. DOI: 10.1029/2004JD005420

[59] Vuille M, Burns SJ, Taylor BL, Cruz FW, Bird BW, Abbott MB, Kanner LC, Cheng H, Novello VF. A review of the south American monsoon history as recorded in stable isotopic proxies over the past two millennia. Climate of the Past. 2012;**8**:1309-1321

[60] Vuille M, Bradley RS, Werner M, Keimig F. 20th century climate change in the tropical Andes: Observations and model results. Climatic Change. 2003;**59**(1-2):75-99

[61] Klein AG, Isacks BL, Bloom AL. Modern and last glacial maximum snowline in Peru and Bolivia: Implications for regional climatic change. Bulletin de l'Institut Français d'Études Andines. 1995;**24**(3):607-617

[62] Ginot P, Schwikowski M, Schotterer U, Gággeler HW, Gallaire R, Pouyaud B. Potential for climate variability reconstruction from Andean glaciochemical records. Annals of Glaciology. 2002;**35**:443-450

Chapter 4

The Unknown Southernmost Glaciers of Europe

Emil Gachev

Additional information is available at the end of the chapter

http://dx.doi.org/10.5772/intechopen.68899

Abstract

This chapter presents the perennial firn/ice patches in the mountains of the Balkan Peninsula. The detailed study of these features in the last decades has proved that many of them are, in fact, small glaciers. They have survived without complete melting since the end of the Little Ice Age, and thus the time of their formation must have not later than in 14–15th century AD. At present, the existence of 16 small glaciers is suggested (and proved for some of them) in three mountains throughout the peninsula: Prokletije (mainly in Albania), Durmitor (in Montenegro) and Pirin (in Bulgaria), the biggest number being found in Prokletije. The two small glaciers (glacierets) in Pirin mountain are at present the southernmost glacial masses in Europe (the only located south of 42°N). Despite the registered warming of high mountain climate, small glaciers on the Balkan Peninsula have shown no trends towards shrinkage for the last 23 years.

Keywords: small glaciers, snow patches, Pirin, Durmitor, Prokletije

1. Introduction

Few mountains in Europe host classical glaciers at present: The Alps, the Great Caucasus range, the Scandinavian mountains, Polar Ural and the Pyrenees [1]. Apart from them, there are numerous small bodies of firn and ice in other mountain ranges across Europe which are still of a permanent character, with their mass moving down by gravity. Of a special interest are those in the mountains of Southern Europe [2]. They represent the furthest glacial outposts, some of which located at almost subtropical latitudes (41–43°N). Most of them exist well below the present climatic snowline, in places of favourable topography and local climate. The marginal conditions in which they still persist, and their great sensibility on short-term climate variations, make them perfect natural indicators and objects for climate change studies. The present chapter will focus on small glaciers on the Balkan Peninsula. Here, at present, the southernmost glacial masses of Europe are located [2] (**Figure 1**).

Mountains with present-day small glaciers

	mountain	*country*	*highest point*	*location*
1	High Tatra	Slovakia, Poland	2655 m	49°12' N 20°13' E
2	Carnic Alps	Austria	2782 m	46°37' N 12°54' E
3	Julian Alps	Italy, Slovenia	2864 m	46°23' N 13°50' E
4	Kamnik Alps	Slovenia	2558 m	46°22' N 14°34' E
5	Gran Sasso	Italy	2912 m	42°28' N 13°34' E
6	Durmitor	Montenegro	2523 m	43°07' N 19°04' E
7	Prokletije	Albania	2694 m	42°26' N 19°49' E
8	Pirin	Bulgaria	2914 m	41°46' N 23°24' E

Classical mountain glaciers

Figure 1. Mountains in Southeastern Europe with present-day small glaciers.

Scientists categorize small sustainable firn and ice features mainly in three types: small cirque glaciers, glacierets and snow patches [3].

Small cirque glaciers and glacierets occupy small sections of Pleistocene glacial cirques (usually just below tall rock walls) and can be considered as remnants of former cirque glaciers, which existed during the termination phases of the Wuermian ice age. On the Balkan Peninsula, these features occupy areas of 0.5–5 ha and have thicknesses in the order of 10–20 m. Moraine ridges have framed their lower ends. Small cirque glaciers have elongated contour, a longitudinal profile with a concave upper part and convex lower section of a tongue shape. Glacierets have simpler longitudinal cross-section (convex or concave or straight), lack of a pronounced tongue-like end and the width is often greater than length [4–6]. The presence of dynamic downward motion of firn-ice mass has been, however, proved in both types [3]. Snow patches on the other hand are not considered glaciers. They are either forms that are not permanent in a long-term sense (for more than several years), or they persist in time, but without conditions for motion. Features of the latter type often occupy karst sinkholes.

On the Balkan Peninsula, the present existence of at least 16 small glaciers has been documented and studied in three of the highest mountains: Pirin (in Bulgaria), Prokletije (mainly in Albania) and Durmitor (in Montenegro) [3–18]. Some specific conditions combine to make them possible to exist 600–800 m below the present position of the climatic snow line (estimated between 2700 m in the Western Balkans and 3200 m in the Eastern Balkans, [12, 19]): (1) altitudes between 1900 and 2600 m a. s. l. that provide for relatively low, annual, and seasonal air and ground temperatures (annual: +1 to +3°C, still too high for glaciers); (2) karstified carbonate bedrock, lightly coloured, with high albedo. It does not get warm too much in summer and allows the drainage of glacier meltwaters during ablation season, thus hindering glacier melt; (3) Shaded locations in former cirques (North or NE aspect, below high rock cliffs); (4) High winter precipitation and great contribution of avalanche and windblown snow in mass accumulation, which allows to effectively double the actual amount of snowfall.

High mountain climate conditions at these altitudes define two seasons in the annual cycle of small glaciers: accumulation season (from November to April) and ablation season (from May to October), and a balance year that can be considered similar to the hydrological year accepted for the region [20]: November 1 to October 31. Autumn (September 15 to October 31) is the best time to observe small glaciers and measure the results of the consecutive mass balance year.

2. Research of small glaciers

A wide variety of methods are applied in the research of small glaciers, many of them are specific. The knowledge about these features can be addressed as 'microglaciology', a field of science that bridges between classical glaciology and periglacial geomorphology.

Mass balance studies reveal inter-annual variations of small glaciers. The most accurate would be to measure changes of firn/ice volume. This is hard to do as it requires detailed knowledge about glacier subsurface topography, and laborious measurements after both the accumulation and ablation season. If small glaciers are to be observed mainly as climatic indicators, it is often enough to know relative changes from year to year and in longer terms. In this context, it is easier to measure the surface area of small glaciers or, as alternative (or in addition), to record fluctuations in glacier front or ice level by a measurement on the field or by photographs. It is desirable that these are done at least once a year, in autumn, to summarize the results of the ending mass-balance cycle (balance year).

On the field, glacier area is measured usually with a measuring tape (or rope) and a laser range finder. Measurements are done along glacier contour (in cases of a simpler shape) or on selected lengths and widths (in cases of irregular contours). Satellite image data can also be used if done in the exact time of the year. Distances of glacier fronts from fixed positions are measured in cases of changeable lower ends, again with the use of a tape. Current positions of firn level can be marked with paint on the rock. Measured data are then processed in a laboratory: measured lines are entered in Geographic Information System (GIS) in an appropriate scale, and then, the software calculates the exact areas. When only selected lengths and widths are measured, they are entered in the software in a proper scale. They are used as a frame, on which photographs of the glacier surface, made from distant positions, are then fitted. After that glacier contour from those images is digitized, and the area is calculated.

Spatial overlay of data from multiple measurements allows for precise comparison between glacier states of different years. Repetitive photography is also an important technique to obtain inter-annual changes. Glaciers are photographed each time from same (fixed) positions, and then images are overlayed. Precise data about area and volume cannot be obtained by using this method, but it is highly indicative when tracing the relative changes and trends in the development of small glaciers. Later, if proper scaling is done on the field, accurate absolute values for surface area and level variations can be retrieved from such photographs.

The current state of snow and firn cover is also quite indicative for the mass balance from the past year, especially for the evaluation of accumulation and ablation varieties across glacier surface. It is assessed on the field, with the use of alpine equipment (crampons, ice axe, etc.).

Effects from accumulation season only are studied in spring (April–May) by measuring snow cover thickness and density in glacier vicinities. Such observations have been rare, especially in our region, due to the high avalanche danger and limited accessibility of glacier sites in that time of the year.

Morphology studies involve geomorphological, glaciological and geophysical methods. Morphology analysis aims to reveal how a glacier is formed. It requires a detailed description of glacier surface geometry (contours, tilts, bergschrund, crevasses, caverns) and the character of surrounding landforms (moraines, protalus ramparts, avalanche gullies, screes). Size and roundness and lichen cover of debris are assessed. Weathering of depositional forms can be examined, e.g. with a Schmidt hammer [21, 22]. However, only relative age can be assessed with these methods.

Internal structure of small glaciers is testified with various techniques, which require investments in labour and equipment. The easier way is to excavate pits in glacier body, but this is hard to do and not much informative as pits cannot be deep. It is better to study natural outcrops of glacier body instead (bergschrund or cracks). Drilling with appropriate ice drills allows to reach depths below 10 m and to retrieve unspoiled cores for analysis in a laboratory. Radar sounding makes possible to estimate underground structure (thickness, sediment layers, patches of buried ice) without digging [23]. Both drilling and sounding, however, require carrying out heavy and expensive equipment, and this sets limitations on the application of these techniques, especially in hardly accessible high mountain areas.

Isotope composition of firn and ice along with absolute ages of formation of glaciers and their surrounding landforms (e.g. moraines) can be verified with the use of laboratory techniques after taking samples of rock, ice or organic particles. Such analyses (isotope, radiocarbon, etc.) are costly and have been applied just for two of the small glaciers on the Balkan Peninsula (in Pirin [4–6] and Durmitor [24, 25]). Absolute ages of surrounding moraines have been also retrieved by lichenometry for a glacier in Durmitor mountains [13].

3. An overview of small glaciers on the Balkan Peninsula and their research

Glacial nature has been already proved for four glaciers: two in Pirin, one in Prokletije and one in Durmitor. Twelve more features in Prokletije mountain are considered as, most probably, glaciers on the basis of their morphology and behaviour in the last decade [3].

3.1. Pirin

Pirin is the second highest mountain in Bulgaria and the third highest on the Balkans (**Figure 2**). It rises in the south-west part of the country, reaching an altitude of 2914 m a. s. l. at its highest point—Mt. Vihren. The mountain is a horst block, oriented NNW to SSE, which is built of granitic intrusions and a mantle of metamorphic rocks. A section of the northern part has on its top a thick (500–1000 m) cover of marble that composes the main ridge and the northern slope. Several gigantic glacial depressions (cirques) were formed in this area during the Wuermian ice age and subsequently have been karstified. At least four sustainable snow/firn features have been discovered and mapped in this area. Two of them have been proved to be small glaciers and, more strictly, glacierets [3, 5, 7].

Snezhnika glacieret is located at 41°46′09″N and 23°24′10″E at 2400–2450 m altitude a. s. l., just below the north-eastern marble wall of Vihren peak. The glacieret has an eastern exposure and a trapezoid shape with length of 80–100 m and width about 90 m. About 4–5 high moraine ridge surrounds glacieret body from three sides. It is considered to be formed in its present shape in the Little Ice Age [4–6] (**Figure 3**).

The first measurements and drilling of Snezhnika were done in 1957–61 by the Bulgarian karstologist Vladimir Popov, in relation to the Third International Geophysical Year [26]. The drilling reached the bottom at 8-m depth. Regular climatic measurements were performed in a meteorology station, which was equipped with a thermograph and gauge for total precipitation [27]. After the end of the programme, research was abandoned. German scientists from Dresden measured glacieret area in the autumns of 1994, 1996 and 1998–2007. In 2006, they made three drillings of the firn, the deepest reaching the bottom at 11 m. Glacier ice with a density of 0.9 Kg/l was found in the cores at depths below 10 m, and the radiocarbon dating of organic particles from these depths confirmed the ice was at least 100 years old [4–6]. This is a direct evidence for the glacial nature of Snezhnika. Since 2008, the glacieret has been also monitored by Bulgarian scientists. Its area has been measured in every autumn for the last years (2008–2016) [3, 19, 20]. After the hot summer of 2012, a cave that is 25 m long and 1.5–2.5 high opened at the bottom, reaching the back wall. There, we observed a cross-section of the glacieret body with sediment strata inside the firn [28]. While going downwards from the highest end (at the bergschrund), those layers changed their tilt from normal to reverse, indicating the presence of slow curvy digging motion, typical for the accumulation zone of mountain glaciers [3, 23].

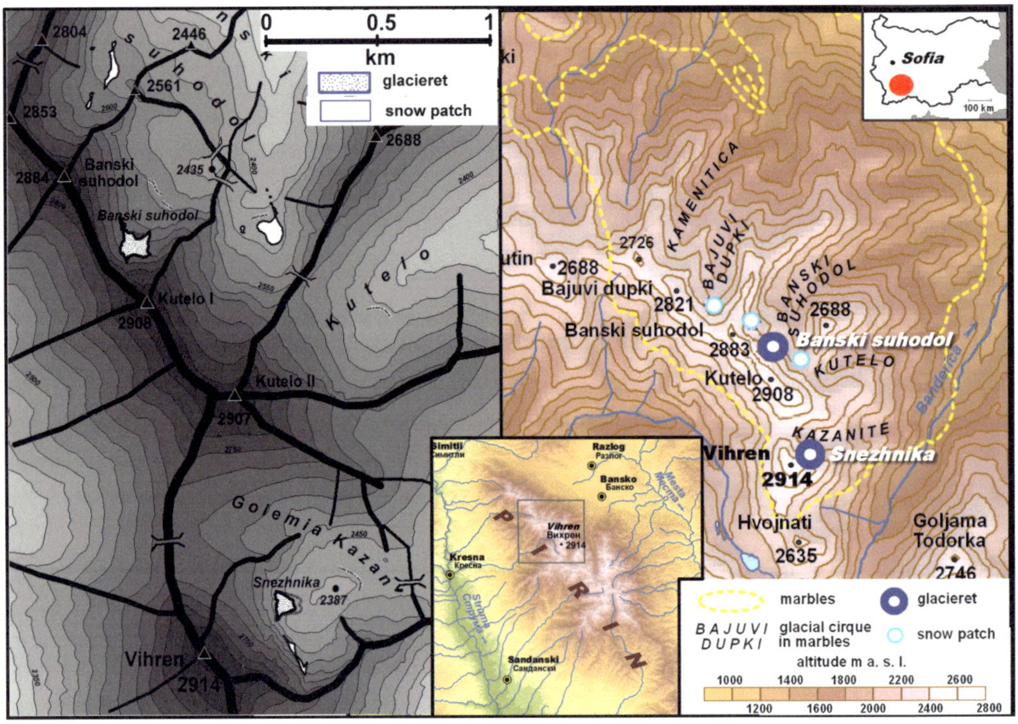

Figure 2. Glacierets and snow patches in Pirin mountains.

In these last 23 years, the area was subjected to large fluctuations, between 0.30 and 0.77 ha, without any specific trend. Average area for the period is 0.55 ha. At present, Snezhnika has been recognized as the southernmost glacial mass of Europe, being, together with Banski suhodol, the other glacieret in Pirin, the only one that is situated in south of the parallel 42°N [2].

Banski suhodol glacieret is situated 1.5 km to the north of Snezhnika, in a vast cirque below the second highest peak in Pirin, Kutelo (2908 m). It has a northerly orientation, irregular shape and an area about 1.2 ha [3, 7, 8]. The altitude of the glacieret is 2610–2700 m. It has a complicated shape, with a length 120–130 m and width 130–135 m. The surface is concave, tilted between 25 and 40°. Two moraine ridges parallel to each other are observed below glacieret front. They are more pronounced in the middle and less on the sides, as avalanche and debris flow paths pass there.

Being hardly accessible, this feature was described and mapped for the first time in 2009 [7, 8] and has been monitored annually since then [3]. Since 2011, the fluctuations of glacieret front have been measured in relation to five fixed points placed on large boulders. The inter-annual fluctuations of the surface of Banski suhodol glacieret are weakly expressed, with a maximum registered in 2010 and minimum in 2012. In October 2012, fresh glacial striations were observed on bedrock surfaces at glacieret front: a direct evidence for glacial type motion of the firn-ice mass of the glacieret.

Apart from the glaciers, two sustainable snow patches are also situated in Banski suhodol cirque. They have been found persistent for the last 8 years, but the low tilts and the closed

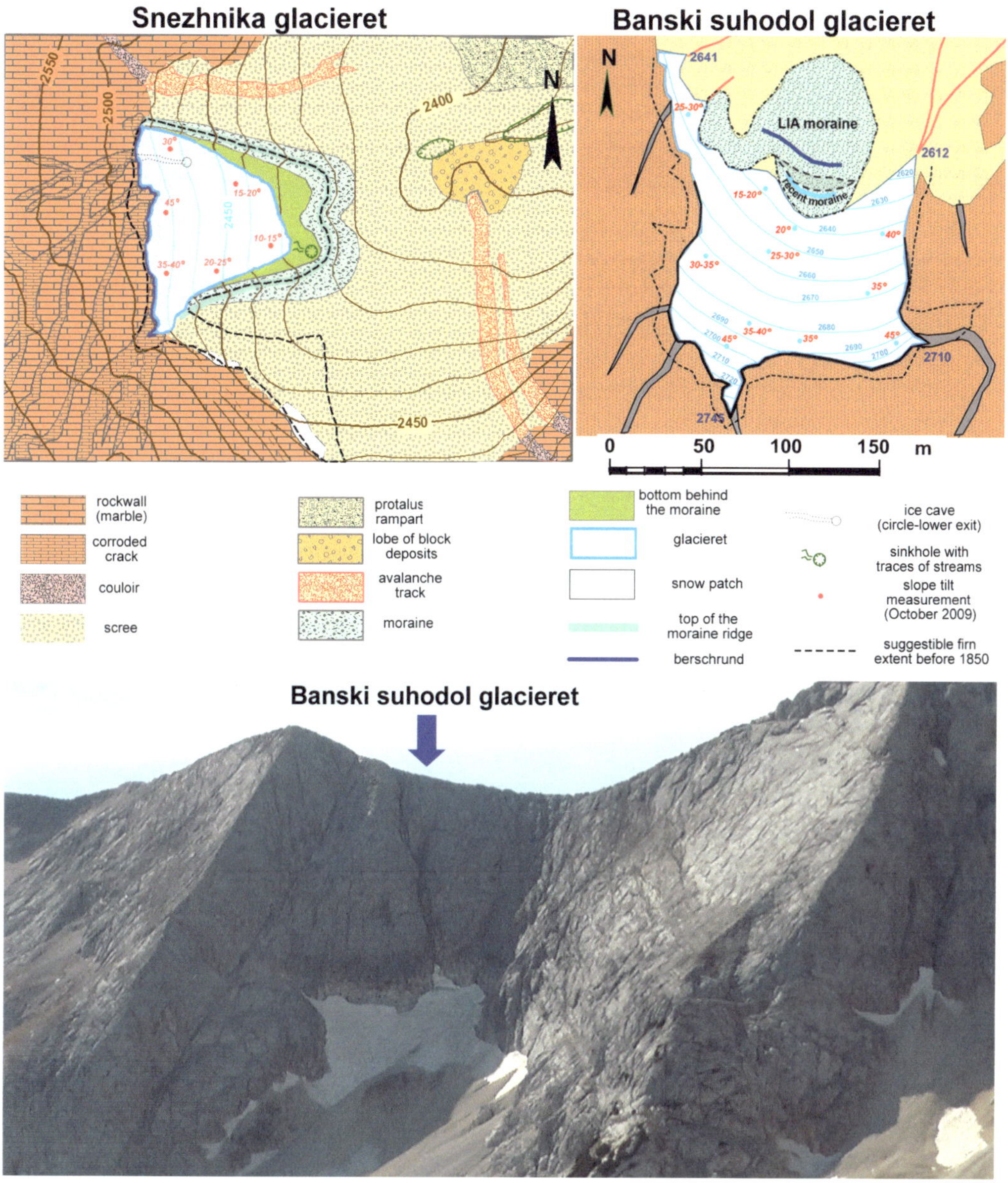

Figure 3. Snezhnika and Banski suhodol glacierets in Pirin.

depressions, where they lie, suggest they can hardly move. Their positions, however, and the moraines that surround them, indicate that they should have been glaciers in the nearly past (maybe in the Little Ice Age).

Presence of a perennial firn and ice was also reported for the cirque Bayuvi dupki, by Hristo Peev in the middle of the twentieth century [29, 30]. He reported about a 500-m long 'firn glacieret' that occupied the bottom of that cirque and gave information about years in which it was greater/smaller for a period of almost 2 decades. Although no figures were given for areas or lengths, this research is considered the first monitoring of a small glacier in Bulgaria. Nowadays, however, no snow remains in this cirque after hot summers. Reports for the existence of sustainable snow patches have been, however, made by enthusiast mountaineers for some locations in the Kamenitica cirque [31].

3.2. Durmitor

Durmitor is located in NW Montenegro. It is the second highest massif in the main Dinaric chain (Mt. Bobotov kuk, 2522 m a. s. l), a very small mountain, situated on a karst plateau at 1450–1550 m a. s. l. close to the deepest canyons of Europe (those of the rivers Tara and Piva [32]). The main part of the mountain is composed of thick Triassic and Jurassic limestones, which to the south overthrust Cretaceous flysch formations [33]. Four vast cirques are heritage from the extensive Wuermian glaciation. In the easternmost of them is the Debeli namet, the only present-day small glacier in Montenegro. The glacier is located on 2030–2200 m altitude; it has a northerly exposure, length about 300–320 m and width 110–135 m [34]. It has a classical elongated contour, with a wider concave upper section and bulged tongue at the front. The glacier is surrounded by a huge moraine, which rises 10–20 m above the surface. An amphitheatre of rocks and couloirs rises more than 300 m to reach the main ridge of the mountain in the south of the glacier: a grassy plateau, at 2400–2450 m a. s. l. (**Figure 4.**). Its lower section represents a surface of barren corroded rocks with a tilt steadier than the rock wall itself. This is the area where Debeli namet expands most, after years, of positive mass balances [9, 34] (in contrast to glacierets in Pirin, which fluctuate mainly in their frontal sections). Tilts of the glacier surface are in the range of 20–25°, reaching 35–40° just at the upper end. Strong mechanical weathering of rocks at the back supplies lots of debris on the ice surface, especially in the SE part.

Debeli namet was recognized as a small glacier by all researchers [2, 3, 9, 13–18]. For the first, it was mentioned in the 1960s [35]. Predrag Djurović from Belgrade, Serbia, measured glacier area in the autumns of 2003, 2006, 2008–2010, and 2015–2016 and reconstructed the size on the basis of aerial photographs for 1961, 1971, and 1981. In 1993, he tried to measure ice velocity with a stick stabbed in the middle part of the glacier. It was found at glacier front after 11 years [18]. Philip Hughes from the UK made size measurements of the glacier in 2003 and 2005–2007 and a lichenometry dating of the surrounding moraine, which addressed its age to the beginning of the twentieth century [13–16]. Accurate surface area measurements of the glacier have been done by Bulgarian scientists every year since 2011 [3]. The ice of the glacier was sampled for heavy metals and radioactive elements [24, 25]. Area observations cover a long period (since 1961) but have become already systematic since 2003. Here, it is also hard to outline any trend in fluctuations that are increasing from year to year (from 1.2 to 3.1 ha, i.e. up to three times). In the last 3 years, however, this glacier has suffered the most dramatic shrinkage on the Balkans. But it is still larger compared to its sizes during the 1990s.

There is one more perennial snow-firn feature in Durmitor, the snow patch in Snežna vrtača, a giant round sinkhole on the plateau of Šlijeme, filled with snow all year round [33].

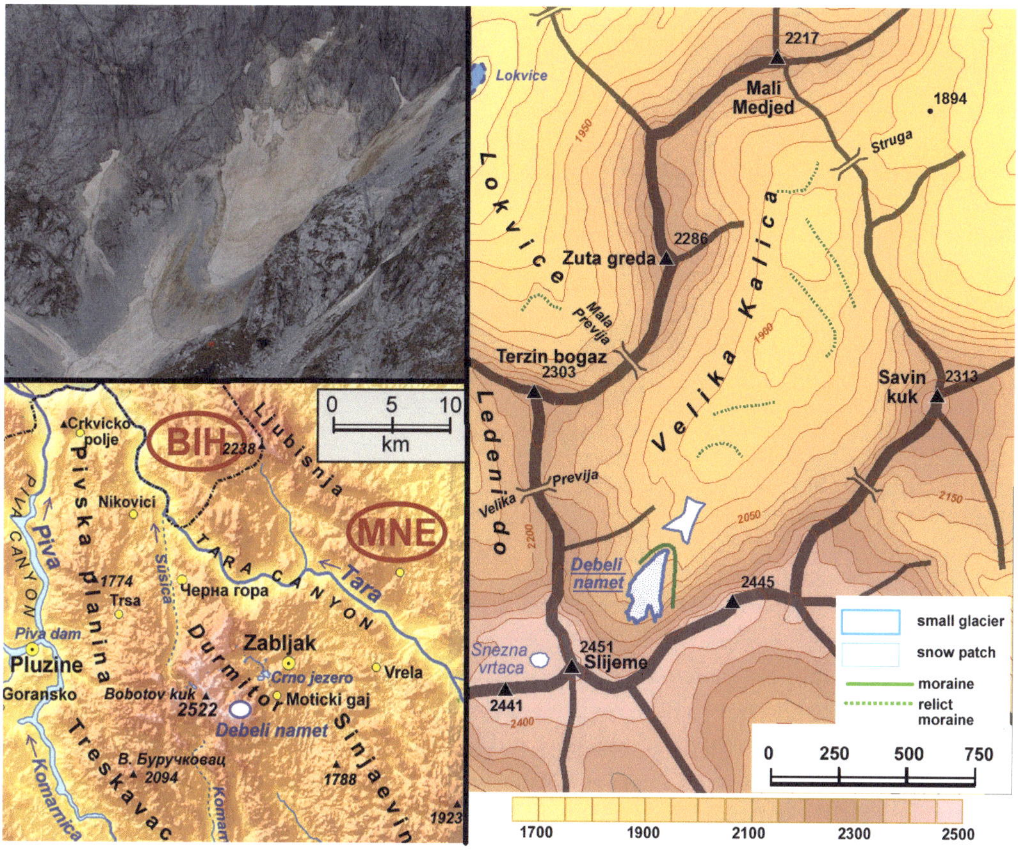

Figure 4. Maps of Durmitor and Debeli namet glacier.

3.3. Prokletije

Prokletije is a large mountain system, situated mainly in the northern Albania, on the borders with Montenegro and Kosovo. It marks the SE conclusion of the main Dinaric chain. Strongly dissected by deep valleys, it rises to almost 2700 m a. s. l. (Mt. Jezerce, the highest of all Dinarides). The central, southern and western sections of the mountain are made of Mesozoic limestone: very thick, tectonically reworked, heavily exerted by Wuermian glaciers and deeply karstified. The present rugged morphology of the mountain reminds of the Dolomites in Italy. In the eastern flanks of the mountain system, silicate rocks prevail and the topography there is smoother and relict glacial relief is much less pronounced (**Figure 5**).

Prokletije is among the least explored mountains in Europe. First, the famous Serbian geographer Jovan Cvijić paid attention to the impressive topography, left from the Pleistocene glaciers in the area around of Plav lake [36]. The presence of perennial snow and ice in the area around the highest point Maja e Jezrecës (Jezerce, 2694 m a. s. l.) was first mentioned by an Austrian topographer, who investigated the area during WWI, and mentioned snow fields more than

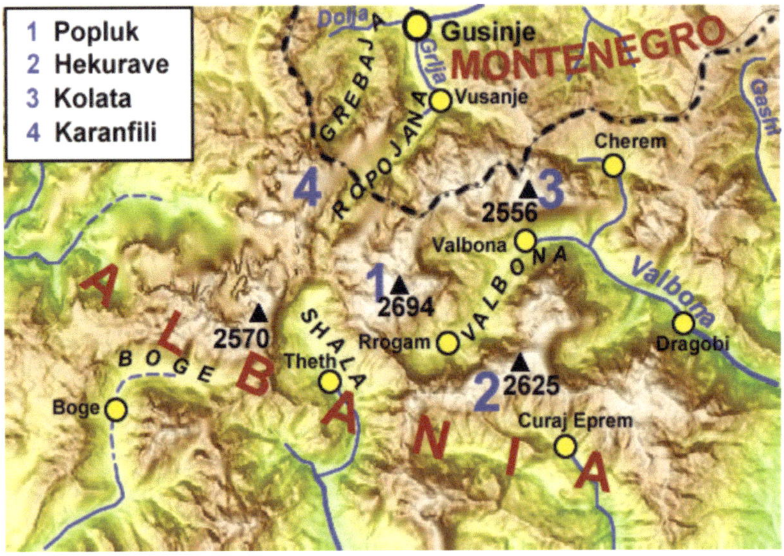

Figure 5. Prokletije mountain with locations of small glaciers.

1 km long [37]. Until the beginning of the twenty first century, geographical studies for this area were very few and not focused on present glaciation (e.g. [38, 39]). In 2007–2008, the area around Mt. Jezerce was researched for relict and present glacial evidence by Serbian geomorphologists, who reported about three 'active glaciers': the largest on the Balkan Peninsula with an area of 5 ha in the cirque Buni i Jezerces at 1980–2100 m a. s. l. and two smaller glaciers to the NE of the highest peak [12]. Soon after, another glacier with an area of 4.9 ha was described by a British expedition to lie under the eastern wall of Mt. Jezerce [15]. Since 2011 the area has been visited by Bulgarian scientists every autumn. As a result, it was revealed that the mentioned feature in the cirque Buni i Jezerces is in fact a snow patch, as it melted almost completely in 2012 and again in 2016. To compensate that, two more small glaciers were declared in the upper part of the same cirque on the basis of morphology. Bulgarian scientists have made several expeditions in other ranges within the carbonate area of Prokletije. In result, a total of 13 suggestible glaciers have been recorded and mapped in four main locations in this range, on altitudes between 2450 and 1910 m a. s. l. [3], but the presence of more is likely as many branches of this extensive mountain system are still unresearched. For the last 6 years, changes in the size of the glaciers and snow patches in the area around Mt. Jezerce have been studied in detail [3, 9–11]. The largest of them, the glacieret Jezerce III, has had an average area of 4.5 ha. Large fluctuation of the firn bodies in terms of surface area was recorded in 2011–2016, with a considerable shrinkage in the years after 2013. However, the observed thickness of some of these small glaciers (15 m and more) indicates that they are still far from complete melt.

3.3.1. Popluk range

Popluk is named the highest part of Prokletije system [38]. It includes Maja e Jezerces peak and the surrounding ridges, separated from the adjacent ranges with clearly defined cols. To the south, Valbona pass (1709 m a. s. l.) makes the transition to the high Hekurave range

(2625 m a. s. l.); to the NE is the Qafa Valbona saddle (2030 m), the pass to Bielić range (Maja e Rosit, 2524 m); to the West is the low Peja Pass (Qafa Pejes, 1690 m) that separates the valleys of Theth and Ropojana and the ranges Popluk and Karanfili. Popluk Mt. Jezerce is surrounded from three sides by large deep cirques: Buni i Jezerces (to the NW), Llugu i Zajave (to the E) and Buni i Gropavet (to the SW) (**Figure 6**).

Llugi i Zajave cirque hosts three small glaciers (glacierets). They are not situated on the bottom of the cirque but on a high terrace just under the 200–300 m high NE rock wall of Mt. Jezerce. Jezerce I (1.2 ha) and Jezerce II (2 ha), located to the NW of the summit point, are typical glacierets: they contact the rock wall, have straight surface and widths larger than lengths. Their fronts, which lie on deeply weathered and corroded limestone blocks, are bordered by moraine ridges several metres high. When they expand, both glaciers join into a single snow field. In periods of retreat Jezerce II disintegrates into several parts. The glacieret Jezerce III is situated further to the SE. It lies on a wide terrace on two levels (at 2400–2450 and 2350–2370 m,

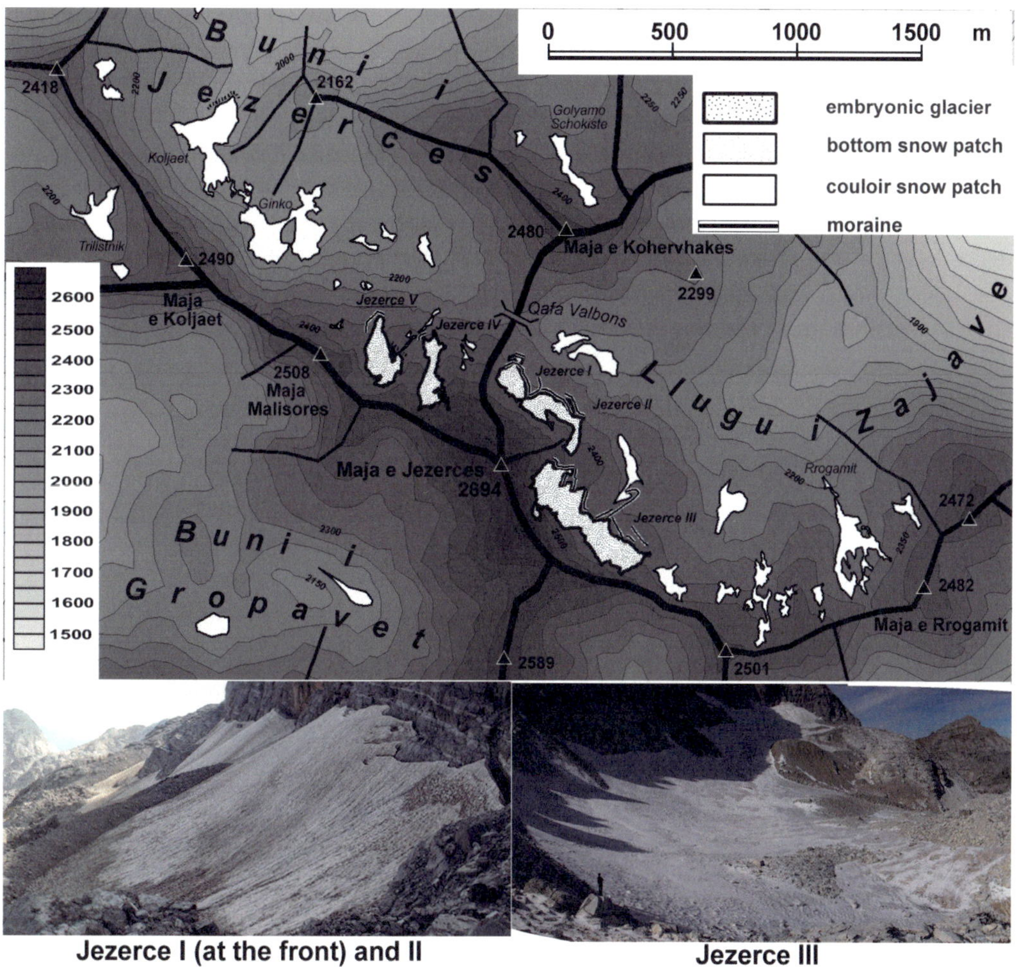

Figure 6. Small glaciers and snow patches in Popluk range.

respectively). However, the firn mass does not actually move down: it is mostly concentrated on the upper level (to the NW), which is on a shadier position, just below the peak. At the time of positive mass balance, all depression is filled with snow, and the glacieret obtains an impressive size (5–7 ha). Jezerce III is the largest small glacier on the Balkan Peninsula. After hot and dry years, ice masses on the lower level defragment to several snow patches occupy sinkhole bottoms, and the glacier becomes limited on the upper level. However, even in such conditions, it remains larger than the others, and ice thickness is still more than 10 m. In October 2014, surfaces of polished rock were observed by us near the firn front to evidence glacial type motion.

Buni i Jezerces is the largest of the three cirques. It is divided into two parts [12]. The lower part is wider oriented to the north. On its grassy bottom, at altitudes between 1750 and 1800 m, there is a group of six glacial karst lakes. Snow and ice features are located mostly in the upper section of the large cirque, which is narrower and oriented to WNW, with altitudes of the bottom between 2000 and 2250 m. Here, glaciers do not lie in the bottom as well but also in deep depressions on a cirque shoulder on 2400–2480 m a. l. The glacieret Jezerce IV (1.8 ha) is in a rocky depression, carved in the NW wall of Mt. Jezerce. It has an irregular shape with length of 270 m and width 70–80 m. The solid rock around produces small quantity of creep material, and in consequence, the two moraines that surround the lower end are tiny. Jezerce V lies further to the west, on the passage between Mt. Jezerce and its western neighbour Maja Malisores (2508 m). It has northern exposure and a pear-like shape of a small cirque glacier, with round upper part and narrow elongated snout. The glacier is situated in a zone of weak rocks, considerable amounts of pebble are produced especially on the SW side, and debris products are deposited as a high moraine ridge on the NW side of the tongue. To the NE of glacier end lies a rock wall, so moraine material is lacking there, and moraines at the very front of the glacier are small, as this area should serve as an avalanche track.

Several sustainable snow patches are found to be spread on the main bottom of the cirque's upper section. Ginko snow patches are in the middle part, at 2100 m a. s. l. They fill bottoms of two sinkholes, lying on a thick cover of scree material. Through years, the snow level can vary by 5–6 m and the area from 0.4 to 4 ha. When the level is high, all patches join in a single one. At the outlet of the upper section is the Koljaet snow patch, which was considered by previous researchers the largest glacier on the Balkans. And indeed, the large snow extent observed in some autumns (up to 4.5 ha for example in 2006 and 2013) and the high moraine ridges at the front can give impression of a small glacier. But regular observations have showed that in other years, the snow was actually missing there. In the small cirque to the North of Maja e Kohervhakes peak, there is another elongated snow field with NW aspect and length reaching more than 200 m in the autumns of some years (e.g. 2013). However, it was completely melted after the summer of 2016, so it is categorized as a snow patch.

The third large cirque, Buni i Gropavet, hosts several snow patches, none of them is considered permanent in long-term sense [12].

3.3.2. *Hekurave range*

This impressive and long range is situated to the south of Valbona valley from its beginning to its end, but the highest part lies to the west of Hekurave peak. This range has a west-east orientation and culminates in the peak Maja Gryk e Hapt (2625 m a. s. l., the third highest in

http://dx.doi.org/10.5772/intechopen.68899

Prokletije). The northern slope is very steep and rocky, and in many regions, it rises almost vertically from Valbona valley.

Up to now, five small glaciers have been discovered in this range [3] (**Figure 7**). Three of them are on a wide terrace in the middle section of the northern slope. The big Glacieret lies at the end of the terrace, in a wide tilted couloir below the northern rock face of Maja e Zhapores (2529 m). It has a trapezoid shape and is bordered by a huge moraine. Next to the east is the Mertur glacier, a small cirque glacier of a classical shape (area around 2 ha), the most representative in Prokletije. It is situated at 2350–2450 m a. s. l. in a zone of weakened rocks

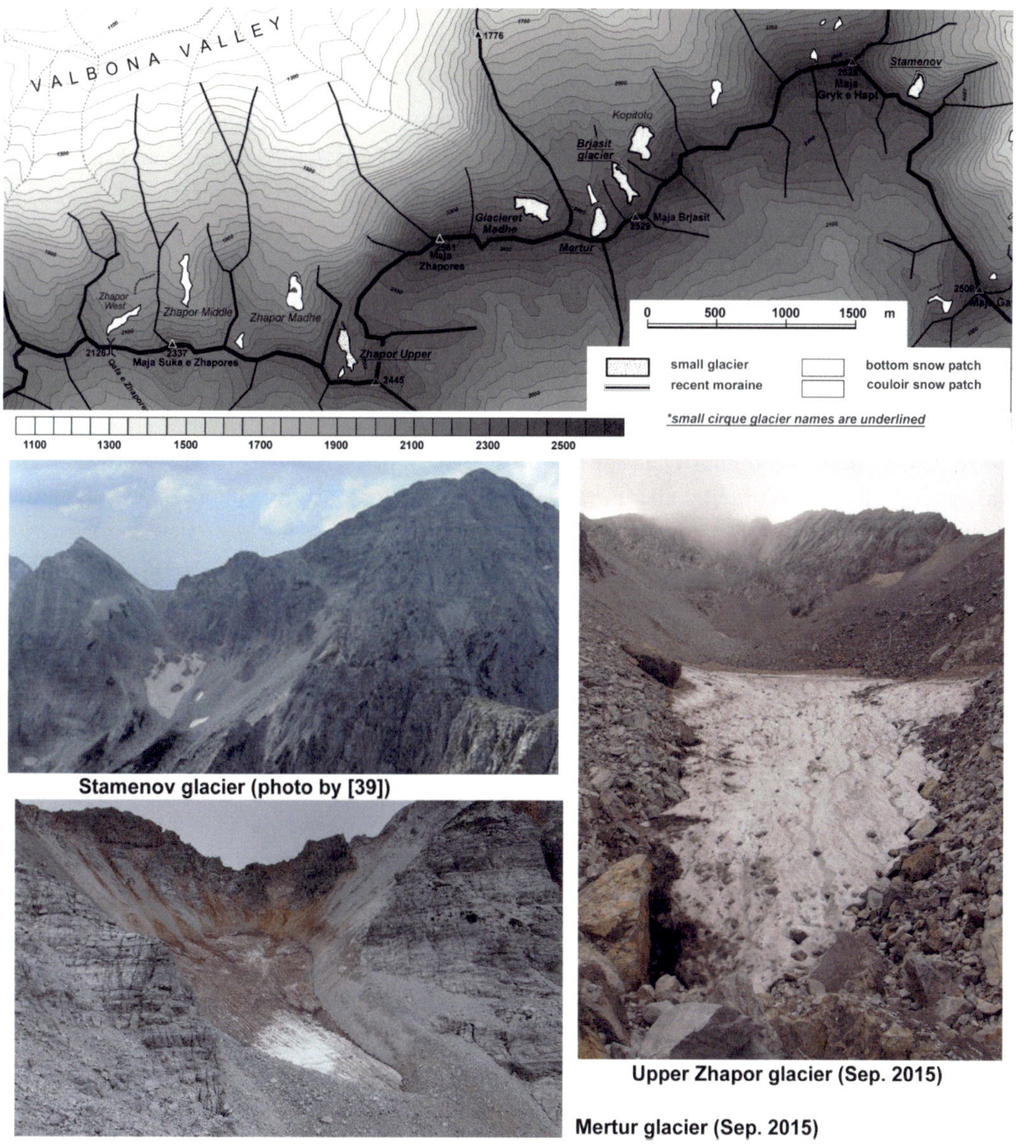

Figure 7. Small glaciers and snow patches in Hekurave range.

(limestone-marble breccia) that cross the ridge in a transverse direction. The scree behind the glacier produces large amounts of debris, which are piled up at the front as a high crescent shaped moraine. Despite that glacier surface is white and clean even after dry summers which indicates the good condition of the glacier and the active recent motion in it. A complex of three fresh stadial moraines is spread down to 500 m from the glacier, indicating its much larger size in the near past. Still further to the NE is the Brjasit small glacier (front at 2280–2300 m a. s. l.), an elongated body of firn and ice (2.5–3 ha), surrounded by moraines from three sides. This feature looks like the one that is made by a giant bulldozer. Further, another firn feature is found in the East. In September 2015, it was all covered by debris. It is obvious that there is buried ice inside, but despite the huge moraine formed behind the front, we accept this feature as a snow patch because of the lack of signs of recent activity.

At least two more small glaciers exist in this mountain range. One of them, Upper Zhapor glacier, occupies a high hanging cirque at 2300–2350 m a. s. l. near Zhapores peak. The glacier has a triangular shape, dictated by the topography of its bed, and in 2015, its surface was relatively fresh, with cracks in the lower part reaching 6–7 m depth. A short moraine made of huge blocks separates glacier end from the beginning of a steep couloir, which descends down to the valley of Valbona. The other feature, Stamenov glacier, lies in an easterly oriented cirque to the East of Maja Gryk e Hapt. It has an elongated shape, a clearly visible moraine that surrounds it from three sides and a relatively fresh look on all images taken in autumn.

3.3.3. Kolata range

This prominent part of Prokletije mountain system lies to the NE of Bjelić range, and is connected to it through the pass Qafa e Presljopit (2039 m a. s. l.). Rising sharply between the valleys of Valbona (to the S), Cherem (a tributary of Valbona, to the N and E) and Zarunica (tributary of Vruja and Lim rivers, to the NW), it is crossed by the Albanian-Montenegrin border and hosts the highest peaks of all Montenegro: Zla Kolata (2534 m a. s. l.) and Dobra Kolata (2528 m). The top of the range is a flattened plateau of flysch rocks, at the eastern end of which rises the highest point Ravna Kolata (2556 m), entirely in Albania. The plateau ends with almost vertical limestone cliffs from all sides. Western and southern slopes are very steep in all their height, descending almost 2 km down to the surrounding valleys. The northern and eastern slopes have staircase profiles. There are several deep and relatively narrow cirques, carved to the north of the plateau surface (**Figure 8**).

Kolata glacieret lies in the deepest cirque with a central position between the three main peaks of this range. The cirque is 200–250 m deep, with vertical walls from three sides, and looks to the north. This glacier has been among the largest and most stable in Prokletije, due to its strongly shaded position, and, possibly, the great contribution from windblown snow from the plateau. It has a triangular shape and minimum observed area about 2 ha. Series of partially developed moraine ridges surrounds it. In years of appropriate conditions, it freely expands to the north, growing to almost 4 ha. Another moraine marks the usual position of the front. Moraines at this glacier are not big, probably due to the solid rock walls that surround it, which are almost lacking wide couloirs. Three smaller features, possibly snow patches, are situated in the other cirques: to the west Malka Kolata snow patch, a remnant of

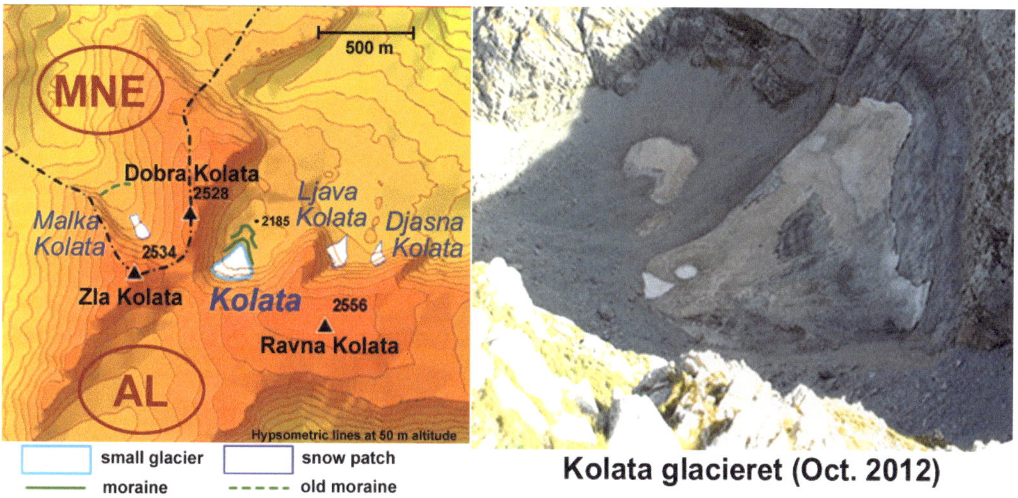

Figure 8. Small glaciers and snow patches in Kolata range.

a small glacier in the past but now looking shallow with no signs of activity; and two smaller patches to the east of the glacieret.

3.3.4. Karanfili

Karanfili range is the NE continuation of the wide Radohima massif that lies to the West of Popluk and Mt. Jezerce. To the east, Radohima share is framed by the deep Ropojana valley and the northern and western numerous ranges fork toward the valley of Vermosh (a tributary of Lim river in Albanian territory). Karanfili ridge goes narrow and sharp between the valleys of Ropojana and Grebaja, crossing the state border between Albania and Montenegro. It contains a number of peaks higher than 2400 m a. s. l., the highest being Veliki vrh (the Great peak, 2490 m) in Montenegro. At the end of Grebaja valley, which is on the NW side and is shorter, the great Grebaja cirque is formed. The two glaciers in this area are found within this cirque (**Figure 9**).

Ropojanski glacier is situated right on the state border line, to the west of the southern peak of the Karanfil (2460 m), and to the NW of the deep Ropojana pass. The altitude of this glacier is 1910–2000 m a. s. l., which makes it the lowermost on the Balkan Peninsula. It has a heart-like shape, with a 4 m high moraine at its front. Another glacier (Switzerland glacieret) has been found to the SW, under the northern wall of Vukoces peak. Framed by rocks from three sides, this feature has created short moraines only on the eastern side of its front. In 2015 the upper part was scattered by stone blocks protruding from the bottom. Several snow patches surround Mt. Vukoces from west and south (in a deep hanging cirque opened to Ropojana valley), but they all were melted in September 2016.

A number of sustainable snow patches are located further to the NE, in deep and very narrow cirques on the Montenegrin side, Kotao and Krošnja. On the bottom of the deepest cirque Kotao, carved north from the three peaks of the Karanfil (North peak, the Great peak, South

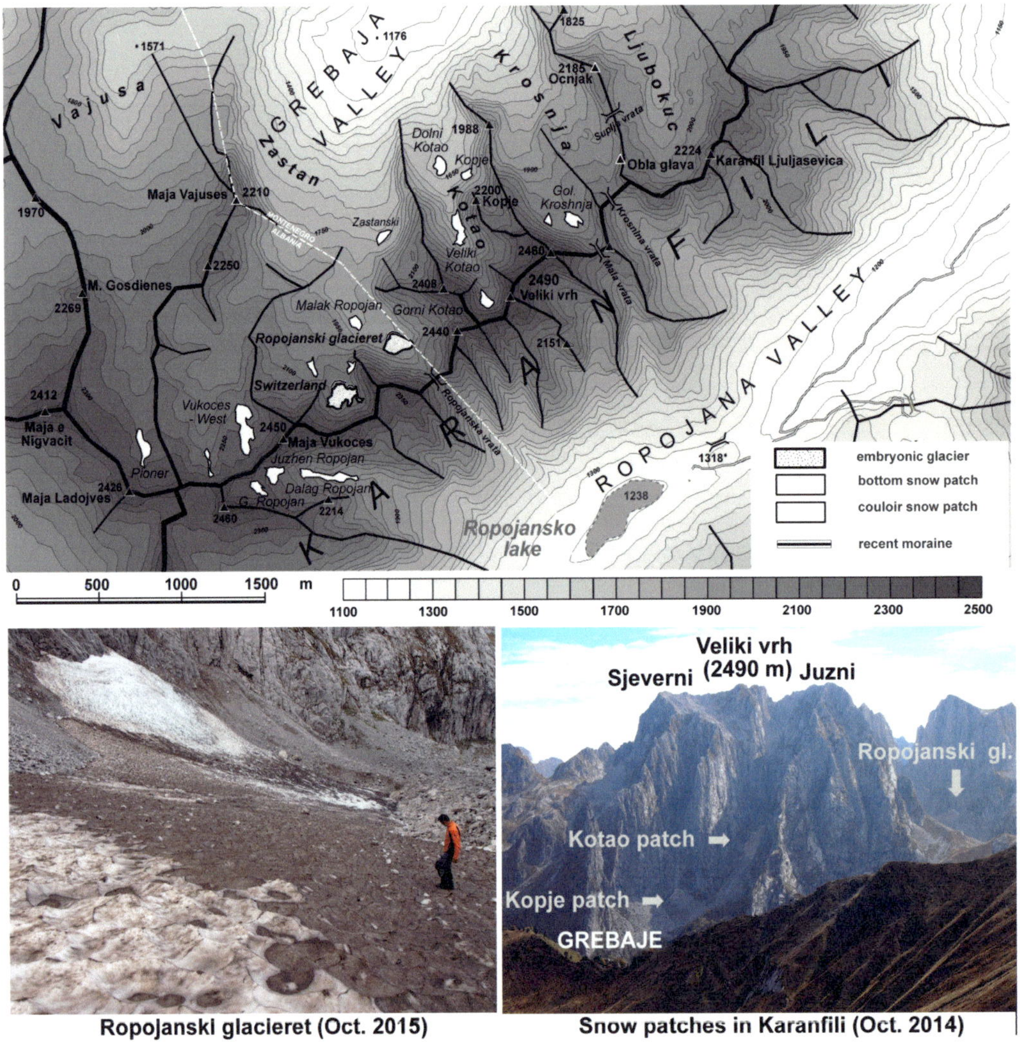

Figure 9. Small glaciers and snow patches in Karanfili range.

peak), are the two lowermost summer-lasting snow patches on the Balkans: at altitudes of 1640 m a. s. l. and 1750–1800 m a. s. l.

4. Revised inventory of small glaciers on the Balkan Peninsula

As a result of all research done by now, 16 sustainable (perennial) firn/ice features in the mountains of the Balkan Peninsula can be indexed in the category of small glaciers (**Table 1**). Summer lasting snow patches have been observed also in other mountain ranges such as in Olympus (Kazania cirque), in Rila (the cirque of the Seven lakes), in Maglić, Korab and others. None of them, however, is proved to have both persistency and indications of dynamic motion.

No.	Name	Type	Mountain	Location (area, cirque)	Co-ordinates		Altitude [m] a. s. l.	Aspect	Length [m]	Width [m]	Area*	
					Lat. N	Long. E					Projec-ted [ha]	Real
1	Jezerce I	Glacieret	Prokletije	Popluk	42°26′42″	19°48′49″	2330–2420	NE	123	147	1.21	1.42
2	Jezerce II	Glacieret	Prokletije	Popluk	42°26′38″	19°48′57″	2330–2445	NE	157	222	2.06	2.61
3	Jezerce III	Glacieret	Prokletije	Popluk	42°26′27″	19° 48′57″	2375–2555	NE	188	271	6.22	7.10
4	Jezerce IV	Glacieret	Prokletije	Popluk	42°26′41″	19°48′33″	2345–2520	N	346	105	1.85	2.30
5	Jezerce V	Cirque glacier	Prokletije	Popluk	42°26′45″	19°48′25″	2330–2435	N	290	153	2.22	2.64
6	Upper Zhapor	Glacieret	Prokletije	Llugu i Silikut	42°23′27″	19°51′38″	2280–2350	N	365	100	1.60	1.70
7	Glacieret Madhe	Glacieret	Prokletije	Llugu i Silikut	42°23′58″	19°52′34″	2250–2380	N	129	263	2.21	2.64
8	Mertur	Cirque glacier	Prokletije	Llugu i Silikut	42°23′55″	19°52′57″	2360–2445	N	213	115	1.55	1.70
9	Brjasit	Cirque glacier	Prokletije	Llugu i Silikut	42°24′04″	19°53′04″	2280–2450	NW	308	84	2.60	2.94
10	Stamenov	Cirque glacier	Prokletije	Llugu i Silikut	42°24′26″	19°54′37″	2120–2270	NE	172	95	1.45	1.64
11	Kolata	Glacieret	Prokletije	Kolata	42°29′00″	19°54′05″	2190–2300	NE	300	195	3.70	4.17
12	Ropojanski	Glacieret	Prokletije	Karanfili	42°29′38″	19°46′43″	1910–2080	NNE	110	156	1.34	1.60
13	Switzerland	Glacieret	Prokletije	Karanfili	42°29′28″	19°46′29″	2130–2225	NNW	142	188	1.68	2.06
14	Debeli namet	Cirque glacier	Durmitor	Velika Kalica	43°07′20″	19°04′30″	2035–2200	NNE	275	145	2.75	3.10
15	Snezhnika	Glacieret	Pirin	Golemia Kazan	41°46′09″	23°24′10″	2400–2445	E	90	95	0.62	0.77
16	Banski suhodol	Glacieret	Pirin	Banski suhodol	41°46′54″	23°23′40″	2610–2700	N	100	127	1.15	1.40

*Area in October 2006.

Table 1. List of small glaciers in the mountains of the Balkan Peninsula.

5. Inter-annual size variations of small glaciers on the Balkan Peninsula

Precise data about size in autumn (at the end of the balance year) have been gathered for Snezhnika glacieret in Pirin for 24 different years, the 21 of which have been consecutive (1996–2016). The area of Banski suhodol glacieret was measured once (in 2009), but its size fluctuations since then are registered by repetitive photographs, and since 2011 the front advances/retreats in relation to fixed points have been recorded. Data for the surface area of Debeli namet are available for the years 1954, 1971, 1981, 1993, 1997, 1998, 2003 and 2005–2016 [3, 14, 18]. On the basis of documents, photographs and measurements information about the size of glaciers and snow patches in Popluk area of Prokletije have been gathered for the years 2006, 2007 and 2011–2016 [3, 12, 15]; for Kolata glacieret: for 2012 and 2014; for glaciers in Hekurave range: for 2006, 2011–2014 [3] and 2015; and for the snow patches in Kotao cirque: for 2006, 2009, 2013, 2015 and 2016 [3, 40].

In general, in short-term small glaciers on the Balkans, size variation of high amplitudes has been demonstrated but differences have been observed in overall amplitude, the way of expansion/shrinkage and the expression of changes. For the whole region, 2005/2006 balance year was a year of glacier growth. For the period of continuous observation in the three mountains, 2010–2014 episode was characterized by synchronous behaviour of all glaciers and snow patches on the Balkans: shrinkage in 2010/2011, 2011/2012 and 2013/2014 and expansion in 2012/2013 balance years. In the next years, different trends were observed in the Eastern and the Western Balkans: for 2014/2015 and 2015/2016 glacierets in Pirin have been stagnating (Snezhnika had even little growth in 2015), while features in Prokletije [41] and Durmitor have been strongly diminished (**Figure 10**).

For this last period, glacierets in Pirin reached their absolute minimum after 2011/2012 balance year, and the size for the years 2014–2016 was similar and at the same time much bigger. In 2015, Debeli namet glacier in Durmitor was smaller, and in 2016, it was much smaller than it was in 2012. In Popluk (Prokletije), 2012 was the minimum for the lowermost snow patches: Koljaet snow patch disappeared almost completely, then reappeared in the next year with a size comparable to that of 2007 when the Serbian scientists had visited it. Shrinkage started again in 2014, continued in 2015 and in 2016, size was again smaller but still a little larger than in 2012. In contrast, for the higher located glaciers in the area, 2015/2016 was the year of the absolute minimum with sizes definitely smaller than those for 2011/2012. For this later period, the maximum size in all the Balkans was registered in 2012/2013 balance year. In Prokletije areas, sizes were larger than those in 2005/2006, Debeli namet glacier was of same size in both the years, and Snezhnika was larger in 2006 than in 2013.

Data for a longer term, available for Snezhnika and Debeli namet, show no trend towards shrink or growth. They both reached absolute minimums in the 1990s of the last century. After 2002–2003, they stabilized and grew, but since 2010, controversial trends have been observed (**Figure 11**).

Figure 10. Inter-annual changes of Snezhnika glacieret (Pirin), Koljaet snow patch (Prokletije) and Debeli namet glacier (Durmitor) for the period 2011–2016.

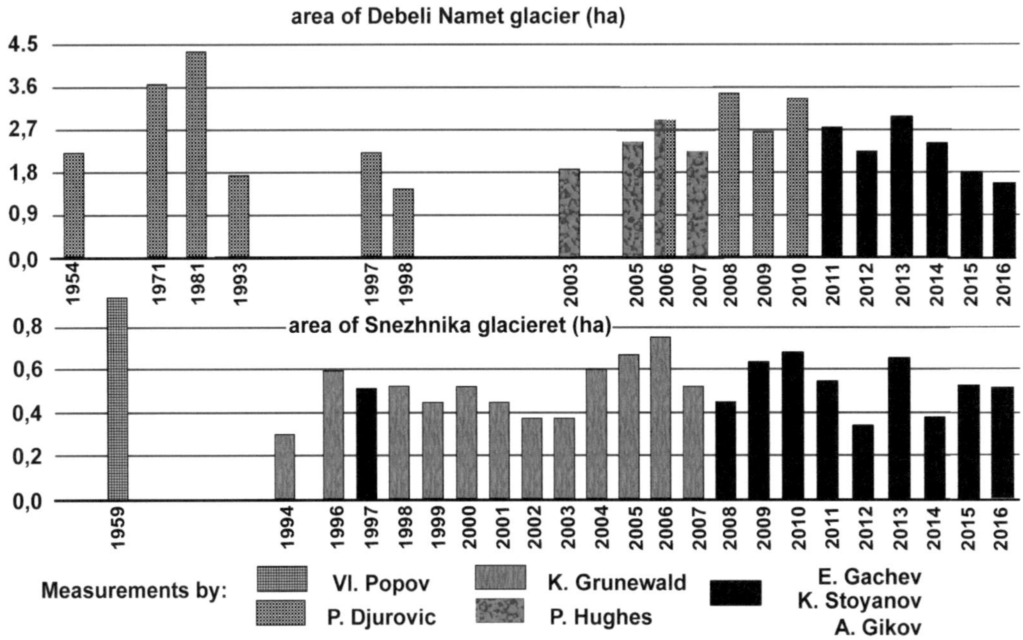

Figure 11. Area measurements for Snezhnika glacieret (Pirin) and Debeli namet glacier (Durmitor).

6. Discussion: small glaciers on the Balkans and climate variations

To understand the nature of short-term glacier variations, we have to bear in mind the climate of the mountains which contain these glaciers. All the three discussed areas are in the zone of transition between the temperate and subtropical (Mediterranean) climate. Located close to the Mediterranean sea (70–100 km away), they are not standing right on the coast, and being among the highest ranges, they are open to continental influences from mainland Europe [3, 12, 33, 42, 43]. Climatic data for these high mountain areas are also lacking. For reference, for a longer period in Pirin, the climatic station of Musala peak in Rila (2925 m a. s. l., 54–55 km away from the glacierets) is used [44]. In the last years measuring devices have been installed in the target area of Pirin such as in Golemia Kazan, close to Snezhnika, by K. Grunewald (an automatic meteorological station recording since September 2011) and on the top of Vihren peak by the South-west University of Bulgaria (logger-sensors, recording every 30 minutes air temperature and humidity since October 2014; and ground temperature since 2016). The statistically significant correlation of temperature data between Musala peak and Golemia Kazan cirque shows that the information from Musala (available also in Internet at [44]) can be used to estimate conditions in Northern Pirin [45]. However, for the last 5 years data from the station near Snezhnika have shown quite high air temperature (annual around +2°C [46]). Analysis of data from Musala enabled to calculate temperature monthly and annual averages for 1994–2016 (**Figure 12**).

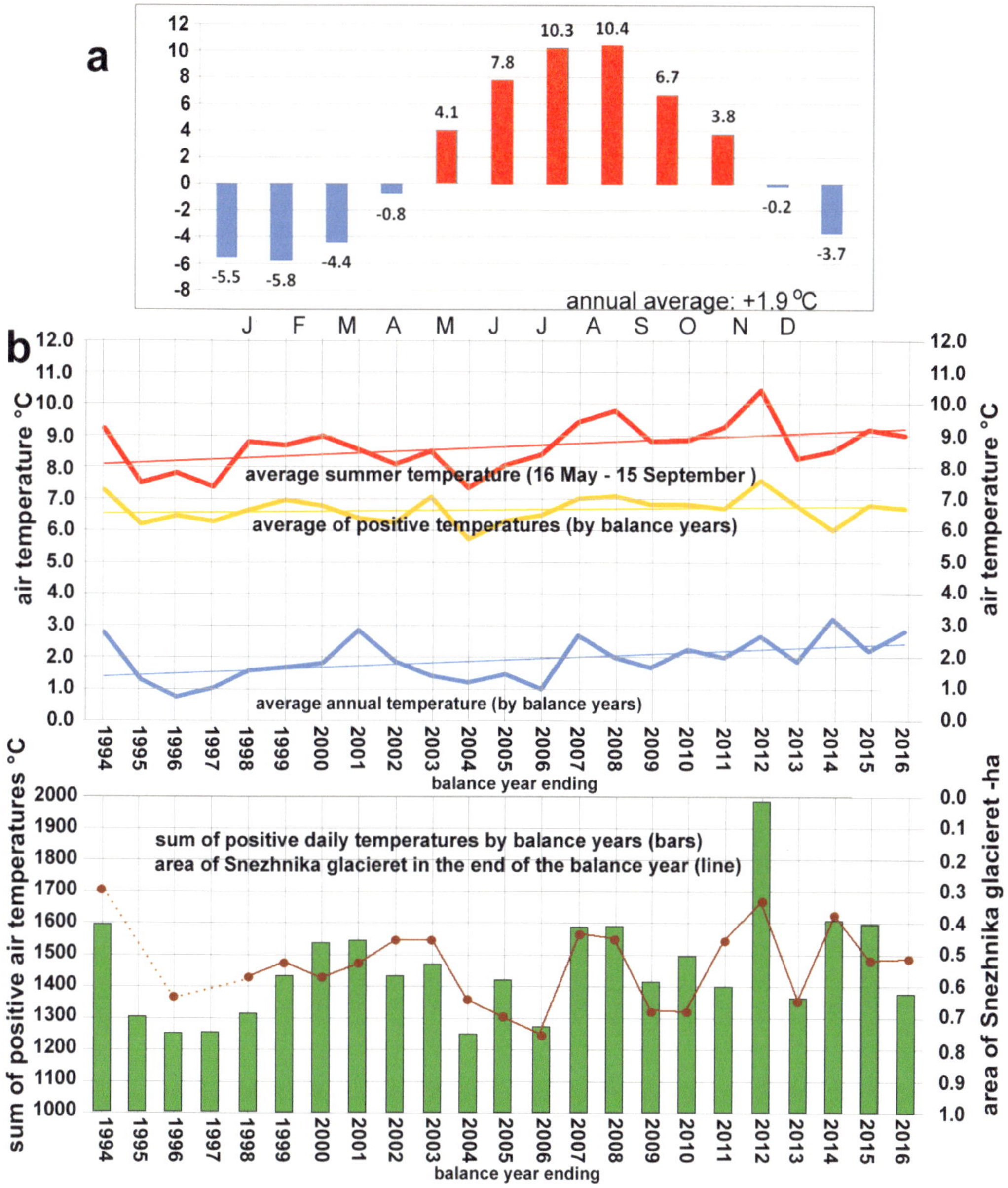

Figure 12. Climatic averages for the area of Snezhnika glacieret based on data from Musala peak: (a) monthly averages for the period 1994–2016; (b) averaged data for 1994/1995—2015/2016 balance years.

What is clear from the figure is that according to the data, there has been registered a considerable warming trend for last 23 years in annual, ablation and summer temperatures. Analyses showed that there is a relatively good correlation between thermal variables, calculated in balance years, and the area of Snezhnika glacieret. For the 23 year period the best is the correlation with the sums of positive temperatures (ablation season sums) which is −0.73.

Why is then no trend in the development of Snezhnika glacieret, if the temperature rise is a fact? The answer is sought in the influence of precipitation, but data about this climatic element are almost missing in Pirin. However, analysis of precipitation data from Musala (with lots of uncertainties) shows that the stagnation of Snezhnika can be due to the higher sums of winter precipitation, which have been registered in most of the years after 2004.

As the annual precipitation in the high parts of Pirin is suggested to be around 1000–1100 mm/year, 650–700 mm of which during the accumulation season [26, 42], Snezhnika is fed to a greatest extent by avalanche and windblown snow. Thus, it receives snow amounts much larger than the actual sum of atmospheric precipitation. On the contrary, Banski suhodol has much smaller avalanche catchment [8]. It relies most of all on shading, and its variations through years are smaller than those of Snezhnika [3]. Sadly, precipitation data from the devices installed in Pirin are not reliable [45, 47].

No climatic data are available from the high mountain areas of Prokletije and Durmitor, the closest mountain station being on Bjelašnica peak in Bosnia and Herzegovina (2067 m a. s. l.). Extrapolations of temperature for the last decade however suggest that around 2150 m a. s. l., annual temperatures are around +2°C and more, and even near the highest glaciers, they are positive [12, 14, 18]. These are however temperatures for open slopes. In negative forms, values are by no doubt lower but still high to sustain glaciers. In the Western Balkans, the existence of perennial ice is favoured by the much greater precipitation: annual amounts for the highest areas of Durmitor are about 2600 mm [18, 33], and for the central and western parts of Prokletije 2500–3300 mm, 2/3 of this amount falling in the cold half of the year [12, 43, 48]. This enables formation of glaciers even at altitudes around 2000 m in strongly shaded sites. The plateau surfaces in the south of Debeli namet and Kolata glaciers serve as great sources of snow, so the actual amount of snow can be more than twice the winter precipitation sum. Glaciers around Mt. Jezerce rely most of all on high altitude (comparable to that of Snezhnika in Pirin) and precipitation around 2500 mm/year and those in Karanfili range mainly on strong shading. Glaciers in Hekurave area, especially Mertur glacier, have always been in good condition in the last years (even in 2012 and 2016). This is due to their high altitude, and, possibly due to much higher precipitation (probably around 3000 mm/year), a result of their southern position and greater proximity to the Adriatic.

The different trends in small glaciers in the Western and the Eastern Balkans, which were observed in the last two balance years (2014/2015 and 2015/2016) can be explained with some synoptic events of accidental character that affected unevenly the territory of the Peninsula. After a relatively snowless winter, in the beginning of March 2015, a cyclone coming from Greece reached Southern Bulgaria and deposited abundant snow in high mountains, triggered avalanches and piled more than 10 m of snow over Snezhnika and smaller but still amount over Banski suhodol (as it is less prone to avalanche). At the same time, mountains in the western part of the Peninsula did not face that cyclone and remained with little snow. After the summer, melt resulted in a positive balance for Snezhnika, a slightly negative for Banski suhodol and a strongly negative for all glaciers in the Western Balkans. Similar situation occurred also after the next winter.

7. Conclusions

At least 16 small glaciers still exist in the mountains of the Balkan Peninsula. Prokletije mountain range provides the best conditions for glacier preservation in the region. Favouring factors for glaciers in the Western Balkans are the high precipitation and the greatest dissection of relief that provides optimal shading conditions. Here, the lowermost small glaciers and snow patches on the Balkans are found. Favouring factors for glacier formation in Pirin are the higher altitude and avalanche occurrence. Small glaciers on the Balkans, which are among the southernmost in Europe, still manage to survive in conditions of climate warming, proved by data from high mountain stations. Further, in a longer term (the last 20 years), they have shown no trend towards shrinkage. Although their area at the end of the balance year shows some correlation with summer temperatures, the neutral balance is reached due to the increased winter precipitation, especially in the last 12–13 years. These facts support the suggestion that small glaciers in such marginal environmental conditions may last much longer than expected.

Acknowledgements

Since 2012, our research of small glaciers has been widely supported by the South-west University "Neofit Rilski", Bulgaria, through the projects: SRP-B3 'New research in the fields of Chemistry, Geography and Environment', 2012; SRP-B4 'Modelling of biological and ecological activities with the aid of chemical and mathematical methods', 2013; SRP-B3 'Mathematical and empirical methods and approaches in chemical research and environmental protection', 2014; SRP-A10 'Complex research of present-day small glaciers on the Balkan peninsula', 2015; RP-B4/16 'Mathematical and computer modelling in natural sciences', 2016. Our work in 2009, 2010 and 2012–2014 was supported by the Bulgarian Academy of Sciences (the National Institute of Geophysics, Geodesy and Geography). Our measurements in Pirin were helped also by prof. Predrag Djurović from the Faculty of Geography of the University of Belgrade, Serbia (2013) and by Brigitte Magori and Flavius Sorin-Sirbu from the West University of Timisoara, Romania. Special thanks to Assoc. Prof. Dr Michail Mihailov from our department, who gave a vital support for our work.

Author details

Emil Gachev

Address all correspondence to: gachev@swu.bg

Faculty of Mathematics and Natural Sciences, South-west University "Neofit Rilski"
Blagoevgrad, Bulgaria

References

[1] Evropa – priroda i istorija (Europe – nature and history). Available at: http://evropa.tova-onova.net/2011/08/26/hidrografia/

[2] Grunewald K, Scheitchauer J. Europe's southernmost glaciers: Response and adaptation to climate change. Glaciology. 2010;**56**:129-142

[3] Gachev E, Stoyanov K, Gikov A. Small glaciers on the Balkan Peninsula: State and changes in the last several years. Quaternary International. 2016;**415**:33-54

[4] Grunewald K, Scheitchauer J. Klima- und Landschaftgeschichte Sudosteuropas. Rekons truktion anhand von Geoarchiven im Piringebirge (Bulgarien). Berlins: Rhombos Verlag; 2008. p. 178

[5] Grunewald K, Scheitchauer J, Gikov A. Mikrolednitsi v Pirinplanina. (Microglaciers in Pirin mountains). Problems of Geography. 2008;**1-2**:97-104

[6] Grunewald K, Scheitchauer J. Landscape Development and Climate Change in Southwest Bulgaria (Pirin mountains). Springer, Berlin; 2011. p. 161

[7] Gachev E, Gikov A, Zlatinova C, Blagoev B. Present state of Bulgarian glacierets. Landform Analysis. 2009;**11**:16-24

[8] Gachev E, Gikov A. Opisanie i parvo izmervane na snezhnika v tsirkusa Banski suhodol. (A description and a first measurement of the glacieret in Banski suhodol cirque.). Problems of Geography, Bulgarian Academy of Sciences. 2010;**3-4**:90-98

[9] Gachev E, Stoyanov K. Present-day perennial firn/ice patches in the mountains of the Western Balkan peninsula. Studia Geomorphologica Carpatho-Balcanica. 2012;**XLVI**: 51-70

[10] Gachev E, Stoyanov K. Present day small glaciers on the Balkan peninsula. In: Fifth International Scientific Conference - FMNS; vol. 5; 2013; Bulgaria: Blagoevgrad. pp. 79-85

[11] Gachev E, Stoyanov K, Nikolova N. Relations between climate, glaciers and lakes in the Popluk massif, Prokletije mountains, Albania. In: International Conference FMNS; 2015; Bulgaria: Blagoevgrad

[12] Milivojević M, Menković L, Ćalić J. Pleistocene glacial relief in the central part of Mt. Prokletije (Albanian Alps). Quaternary International. 2008;**190**:112-122

[13] Hughes Ph. Recent behaviour of the Debeli Namet glacier, Durmitor, Montenegro. Earth Surface Processes and Landforms. 2007;**32**:1593-1602

[14] Hughes Ph. Response of a Montenegro glacier to extreme heatwaves in 2003 and 2007. Geografiska Annaler. 2008;**90A**(4):259-267

[15] Hughes Ph. Twenty-first century glaciers in the Prokletije mountains, Albania. Arctic, Antarctic and Alpine Research. 2009;**41**:455-459

[16] Hughes Ph, Woodward J. Timing of glaciation in the Mediterranean mountains during the last cold stage. Journal of Quaternary Science. 2008;**23**:575-588

[17] Djurović P. Natural and anthropogenic influences as threats to pits in the Debeli namet glacier, Durmitor National park, Montenegro. Contribucion del estudio cientifico de las cavidades karsticas al conocimiento geologico. Malaga; 1999

[18] Djurović P. The Debeli Namet glacier from the second half of the 20th century to the present. Acta geographica slovenica. 2012; **55**

[19] Gachev E. Inter-annual size variations of Snezhnika glacieret (the Pirin mountains, Bulgaria) in the last ten years. Studia Geomorphologica Carpatho-Balcanica. 2011;**XLV**: 47-68

[20] Gachev E. Klimatichni faktori za mezhdugodishnata dinamika na mikrolednik Snezhnika v Pirin (Climatic factors for the inter-annual dynamics of Snezhnika glacieret in Pirin). In: Conference of Bulgarian Geographic Society; 2016; Bulgaris: Vurshets

[21] Urdea P. Absolute ages, relative ages and weathering in glaciated landscape of Retezat mountains (Romania). Preliminary results. Analele Universitatii de Vest din Timisoara, Geografie. 2006;**XVI**:45-60

[22] Aa A, Sjastad J. Schmidt hammer age evaluation of the moraine sequence in front of Boyarbreen, western Norway. Norsk Geologisk Tiddskrift. 2000;**80**:27-32

[23] Gądek B, Kotyrba A. Struktura wewnętrzna Lodowczyka Mięguszowieckiego (Tatry) w śwetle wyników badań georadarowych (The internal structure of Mięguszowiecki Glacieret (Tatra mountains, southern Poland) in the light of the results of georadar investigations). Przegląd Geologiczny. 2003;**51**:1044-1047

[24] Vasilievič. Sadržaj metala i radioaktivnost leda sa ostataka lednika Debeli Namet na Durmitoru. In: Zbornik radova: Rizik tehnoloških sistema i životna sredina. Nis; 1997. pp. 145-148

[25] Kern Z, Suranyi G, Molnar M, Nagy B, Balogh D. Investigation of natural perennial ice deposits of Durmitor Mts., Montenegro. In: Procedings of the Second International Workshop on ice caves; 8-12th May 2006; Demanska Dolina, Slovak Republic

[26] Popov V. Nabljudenia varhu snezhnika v tsirkisa Golemia Kazan – Pirin planina (Observations on the glacieret in the cirque Golemia Kazan, the Pirin mountains). Izvestia na Geografskia institut, BAN. 1964;**VIII**:198-207

[27] Popov V. Morfologia na tsirkusa Golemia Kazan – Pirin planina (Morphology of the cirque Golemia Kazan – Pirin mountains). Izvestia na Geografskia institut, BAN. 1962;**VI**:86-99

[28] Gachev E. Ima li lednitsi na Balkanskia poluostrov (Are there glaciers on the Balkan Peninsula). Priroda. 2014;**1**:96-103

[29] Peev H. Za niakoi savremenni lednikovi formi v Pirin planina (About some present-day glacial features in the Pirin mountain). Priroda (Nature). 1960;1

[30] Peev H. Prinos kam vaprosa za obrazuvaneto na savremenni lednikovi formi v Pirin (A contribution to the issue about the formation of present-day glacial features in Pirin). Hidrologia i meteorologia (Hydrology and meteorology). 1961;**3**:22-26

[31] Planinarski forum. Available at http://planina.e-psylon.net/viewtopic.php?t=9112

[32] Cerović B. Durmitor i kanjon Tare. Vodić. Beograd; 1991. p. 160

[33] Djurović, P. High mountain karst on Durmitor. Beograd; 2011. p. 206

[34] Gachev E, Sotyanov K. Malkiat lednik Debeli namet v masiva Durmitor, Dinarski planini (The small glacier Debeli namet in Durmitor massif, Dinarides). Problems of Geography. 3.4, 2013

[35] Nicod J. Premières recherches de morphologie karstique dans le massif du Durmitor. Mediterrannée 3. Aix-en-Provence, 1968

[36] Cvijić J. Ledeno doba u Prokletijama i okolnim planinama (the Ice age at Prokletije and the surrounding mountains).Glas srpske kraljevske akademije II. 1913. p. 93

[37] Roth von Telegd K. Das albanisch-montenegrinische Grenzgebiet bei Plav (Mit besonderer Berücksichtigung der Glazialspuren). (Albanian-Montenegrian border region of Plav.). In: Beiträge zur Geologie von Albanien, Neues Jahrbuch für Mineralogie 1, Nowack E (ed.). Schweizerbart: Stuttgart; 1923. pp. 422-494

[38] Palmentola G, Baboci K, Gruda GJ, Zito G. A note on rock glaciers in the Albanian Alps. Permafrost and Periglacial Processes. 1995;**6**:251-257

[39] Milivojević M, Kovacević-Majkić J. Glacijalna jezera Buni i Jezerce – Albanija (Glacial lakes Buni i Jezerce in Albania). Glas srpskog geografskog drustva, LXXXV – 1, Belgrade, 2005

[40] Mulić R. The procletian mountains of Plav and Gusinje municipalities. 40 mountain trails. PD Karanfil, Gusinje; 2009. p. 214

[41] Planinarski forum. Available at http://planina.e-psylon.net/viewtopic.php?t=8172

[42] Velev St. Klimatat na Bulgaria (The climate of Bulgaria). Heron Pres, Sofia; 2010. p. 190

[43] Gjeografia fizike e Shqipërisë, Akademia e Shkencave e RPSSH,Tirana. 1990

[44] Stringmeteo. Bulgarian website fordiscussing weather. Available at: www.stringmeteo.com

[45] Grunewald, K., The station 2 in Pirin – Kazana shelter, data sheet 2013; 2011-2013

[46] Nojarov P, Gachev E. Klimatichni faktori za izmeneniata v razmerite na snezhnika v tsirkusa Golemia Kazan, Pirin planina (Climatic factors for the changes in the dimensions of the snow patch in the cirque Golemia Kazan, Pirin mountain). In: Conference: The science, education and art in the 21st century; vol. 2; 2007; Blagoevgrad. pp. 399-405

[47] Grunewald K, Gachev E, Kast G, Nojarov P, Panayotov M. Meteorological observations in the National park "Pirin". IÖR Dresden, 2016

[48] Stoyanov K, Gachev E. Climatic controls over the recent development of small glaciers on the Balkan peninsula. In: International Conference FMNS, Blagoevgrad, vol. 5; 2013. pp. 86-93

Section 2

Glacial Ecosystems

Chapter 5

The Role of Microbial Ecology in Glacier Retreat

Eva Garcia-Lopez and Cristina Cid

Additional information is available at the end of the chapter

http://dx.doi.org/10.5772/intechopen.69097

Abstract

Glaciers have been considered too hostile to harbor life for a long time. However, they are now recognized as unique biomes dominated by microbial communities which maintain active biochemical routes. Microbial communities inhabiting glaciers are diverse depending on the type of glacier and the area studied. Some glaciers have a marine margin and finish in a calving front, with partly or completely temperate tidewater tongues, this establishes important differences with respect to glaciers with a land margin. Depending on the glacier area studied, microorganisms are also characteristic as they establish a vertical food chain, from the surface photosynthesizers in upper illuminated layers to heterotrophs confined in the inner part. Glaciers are retreating in many areas of the world due to global warming. Microorganisms are their most abundant and unknown occupants. They play a main role, carrying out key processes in the development of soil and facilitating plant colonization when glaciers have ultimately retired. These microorganisms are perfectly adapted to their harsh environment and are very susceptible to environmental changes. This chapter summarizes the role of microbial ecology as indicator of the conservation status of glaciers.

Keywords: global warming, ecosystem, microorganism, extremophile, psychrophile

1. Introduction

Earth's cold environments have been considered uninhabited for a long time. Icy deserts seemed too hostile to harbor life (**Figure 1**). However, glaciers and ice sheets are unique biomes dominated by microbial communities which maintain active biochemical routes [1].

Glaciers are dominated by very specific environmental characteristics such as the low temperatures associated with precipitation in the form of snow, the exposure to the intense wind, and the extreme solar radiation. In summer, there are processes of melting and sublimation

Figure 1. Ecosystems in cold environments. (A) Polar deserts, (B) Glaciers, (C) Icy seas and (D) Icebergs.

that are contrasted with the accumulation of ice in winter, producing an imperceptible but constant dynamism. This is the basis of the life of many of the organisms that inhabit them [2].

Microbial communities in glaciers are different depending on the type of glacier and the area studied. Thanks to DNA sequencing methods, a lot of information about their biodiversity and ecology has been acquired. Firstly, glaciers are of various kinds [3, 4] (**Figure 2**). Some of them have a marine margin and finish in a calving front (**Figure 2(C)**), this establishes important differences with respect to glaciers with a land margin (**Figures 2(A), (B)**). In glaciers ending on land, there is continuous permafrost at ice front (**Figure 2(A)**), while calving glaciers present partly or completely temperate tidewater tongues [4]. Secondly, the growth areas of the glacier (accumulation area) are oligotrophic media for microorganisms. They establish a vertical food chain, from surface photosynthesizers in the upper illuminated layers to protists and bacteria confined in the inner part [5]. These microorganisms are greatly influenced by the melting of surface layers. The diversity of microorganisms in the areas of regression of glaciers (ablation zone and glacial lake) can be lower than in the accumulation area [6], although they are usually more abundant. Predatory species are numerous in these areas, so microbial

Figure 2. Types of glaciers. (A) Gébroulaz glacier ending on land. (B) Literola glacier at Pyrenees, ending on a lake and river. (C) Marine glacier at Livingston Island, South Shetland Antarctica.

diversity decreases. At last, taking into account the horizontal stratification of glaciers, they can be divided into three ecosystems: supraglacial, englacial, and subglacial. Additionally, there has been an increasing interest in characterizing retreating ice fronts of deglaciated forefields with the aim of getting to know how both the richness and the abundance of microorganisms vary in a glacier due to climate change [6–8]. In forefields, mixed communities are observed, whose composition changes very quickly.

2. Earth is a cold planet

From a biological perspective, Earth is a cold planet. Most of the Earth's surface is covered by oceans where temperatures are below 5°C [9], and 80% of the terrestrial biosphere is permanently frozen [10]. Some examples of these cold environments are upper atmosphere, benthic marine zones, polar deserts (**Figure 1(A)**), glaciers (**Figure 1(B)**), subglacial lakes, and icy seas (**Figure 1(C), (D)**) [11].

Snow and ice cover over 108 km^2 of the Earth's surface. Snow in winter can cover up to 12% of the Earth's surface [12], and approximately 10% of the planet's land surface is covered by glacial ice in the form of ice caps, ice sheets, or glaciers, accumulating 75% of the world's fresh water [13]. Mean temperatures observed in snow and ice environments can be highly variable at different depths, sites, or seasons. For example, surfaces exposed to wind are influenced by air temperature. Temperature can range from −50 to −70°C during the winter in the Arctic and Antarctic, respectively, to 0°C in summer [14].

3. Glaciers as biomes

Among cold environments, glaciers are considered biomes that should be recognized as such in their own right [1, 2]. A great diversity of microorganisms belonging to the three main domains (Bacteria, Eucarya and Archaea) has been discovered inhabiting these cold environments. Most of the microorganisms isolated from cold environments are psychotolerant (also called psychrotrophs) and psychrophiles. Psychotolerant organisms can grow at temperatures close to 0°C but have their optimum growth temperature at about 20°C. However, psychrophiles have their optimal growth temperature at 15°C or less [15].

Glaciers are inhabited by microorganisms which maintain active biochemical processes.

To grow efficiently at low temperatures, microorganisms have developed complex structural and functional strategies for their adaptation [16]. The study of these adaptation strategies aims to identify the limits of life at these temperatures. Adaptations include the production of psychrophilic enzymes that are functional at low temperatures with a high catalytic efficiency; the incorporation of unsaturated fatty acids in the cell membrane to improve its fluidity; the synthesis of certain proteins that allow synthesizing others at low temperatures [17]; and the production of compounds that allow the cell to protect itself from frostbite (e.g. sugars, extracellular polysaccharides, antifreeze proteins) [18, 19].

4. Microbial ecology in glaciers

Considering the horizontal stratification of glaciers, they can be divided into three parts: the supraglacial ecosystem, the subglacial system, and the englacial ecosystem [2, 5] (**Figure 3**). These three ecosystems differ in terms of their solar radiation, water content, nutrient abundance, and redox potential [2]. These factors greatly determine the biogeochemical cycles, the type of metabolism, and the diversity and abundance of microbial populations inhabiting glaciers.

4.1. The supraglacial ecosystem

The main habitats in the supraglacial ecosystem are the snowpack, cryoconite holes, supraglacial streams, and moraines. On the glacier surface, the absorption of solar radiation by dark organic matter causes snow and ice melting yielding liquid water that is necessary for microorganisms. Meltwater dissolves nutrients from adjacent debris and even directly from the atmosphere [20].

The sunlit and oxygenated supraglacial surface are populated by autotrophic microorganisms such as microalgae and diatoms; by chemolitotrophic bacteria, which feed on inorganic sand particles; and by heterotrophic bacteria and microeukaryotes.

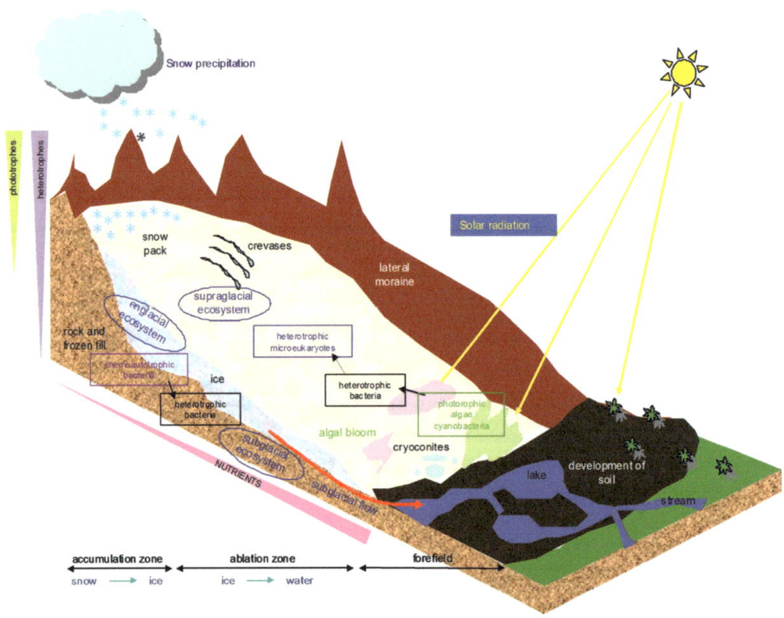

Figure 3. A schematic of different habitats of a glacier colonized by microorganisms. Supraglacial ecosystem is dominated by phototrophic algae and cyanobacteria that take advantage of sunlight and by heterotrophic bacteria and microeukaryotes that feed on organic particles from atmospheric deposition. Microorganisms in englacial ecosystem can be chemoautotrophs, but they can also be heterotrophic bacteria that feed of solubilized products. Subglacial ecosystem is dominated by microorganisms which obtain energy from inorganic compounds and occupy basal ice/till veins and subglacial lakes.

The lithotrophic microorganisms degrade the till and black carbon on the surface of the glaciers. So, the concentration of dissolved ions in water increases and its melting point decreases. This fact develops cryoconite holes, vertical cylindrical melt holes in a glacier surface, which have a thin layer of sediment at the bottom and are filled with water [21]. The materials comprising cryoconite can be divided into two main types: organic and inorganic [22]. Organic matter includes living and dead microorganisms and their products of decomposition, while inorganic matter in cryoconites is dominated by mineral fragments, mainly silicates [22]. Cryoconites are an important microbial habitat in supraglacial ecosystems [1]. Cryoconite holes may converge and origin small streams of liquid water that run downhill [21]. Food webs in cryoconite are dominated by photoautotrophs, mainly cyanobacteria, which provide substrate for heterotrophic communities from a wide range of bacteria. All major groups of heterotrophic bacteria and many fungal groups are represented in cryoconite holes [5]. In addition, microbial eukaryotes such as ciliates are crucial for nutrient recycling through the metabolism of primary producers [23]. Heterotrophic activities in supraglacial habitats are substantial but typically occur at lower rates than the rates of photosynthetic production, which leads to the accumulation of organic matter over time.

Microorganisms inhabiting glacial surface produce a wide diversity of pigmented molecules, which allow their adaptation to cold conditions and solar radiation. They use pigments to obtain energy [24], develop photosynthesis [25], stress resistance [26], and for ultraviolet light protection [27, 28]. For instance, green snow is caused by young, trophic stages of snow algae, whereby more mature and carotenoid-rich resting stages result in all shades of red snow [29]. Dominant species on snow fields belong to the unicellular Chlamydomonaceae. Additionally, some examples of cold-adapted bacteria that produce pigments are the bacterium *Sphingobacterium antarcticus*, which produces zeaxanthin, b-cryptoxanthin, and b-carotene [30]. Other examples include the polar bacteria *Octadecabacter arcticus* and *Octadecabacter antarcticus*, producers of xanthorhodopsin [31], and *Shewanella frigidimarina* which produces the red cytochrome c3 [32, 33]. Colored melanized fungi also live on glaciers, for instance, the oligotrophic genus *Cladosporium* [34]. These pigments absorb solar light and heat, melting snow on glacial surfaces. Microorganisms on glacial surfaces also bear high solar radiation, but in a way, this radiation is beneficial for them. In spring, light radiation melts the glacial surface and leads to the increase in wet areas and the dilution of solutes on snow and ice surfaces, which facilitates the growth of microbial mats [14].

4.2. The englacial ecosystem

The englacial ecosystem presents a minor impact upon nutrient dynamics [2]. Surface meltwaters flood the englacial sediments by means of drainage channels. In englacial ecosystems, live motile bacteria that can reach more than 3000 m of depth. These bacteria live at grain boundaries and other interstices. Mineral substrates such as clay particles [35] provide nutrients and a supply of water for microorganisms. Microorganisms can also live in narrow veins between ice crystals. When the water freezes, dissolved and particulate impurities (including microorganisms) are excluded from the ice matrix into interstitial aqueous channels at the ice-grain boundaries [11]. In turn, these microorganisms and impurities diminish the growth of ice crystals and even break them, facilitating the existence of liquid water. The liquid vein habitat provides water, energy, and nutrients. In contrast with this, the metabolism of microbes encased in solid ice must overcome the diffusion of nutrients in a solid media [36].

Microorganisms in englacial ecosystems can be chemoautotrophs, but they can also be heterotrophic bacteria that feed on solubilized products from pollen grains, invertebrates, and other microorganisms. At great depth, anaerobic respiration can take place [35, 37], and methanogens could be active [2].

4.3. The subglacial ecosystem

At glacial sediments and bedrock, debris contains minerals and sedimentary organic carbon that, combined with subglacial water, create microniches where microorganisms can live [5]. A strong coupling is likely to exist between the hydraulic conditions at the glacier bed and the bacterial processes that take place [20]. The subglacial system is dominated by aerobic/anaerobic bacteria and probably viruses in basal bedrock and subglacial lakes. It also contains diverse, metabolically active archaeal, bacterial and fungal species [38]; although eukaryotes have not been detected in all subglacial environments examined [5].

As there is no sunlight, chemoautotrophic or chemolithotrophic bacteria obtain energy from inorganic compounds. The inorganic processes associated with chemoautotrophs and chemolithotrophs may make these bacteria one of the most important sources of weathering and erosion of rocks on Earth [39].

5. Glacier retreat

Glaciers are highly sensitive indicators of past and present climate change. Their current area and volume are a response to changes in both temperature and precipitation [13], as glaciers respond to slight but prolonged changes in climate. The study of glacier fluctuations is relevant to provide an understanding of climate change over temporal scales [13].

Most glaciers are currently retreating. According to the National Snow and Ice Data Center (NSIDC), the total glacier loss per year since 1994 is approximate 400 billion tons [40–46]. The retreat of glaciers is particularly concerning, since it represents hazards for human communities living near them such as outburst floods, landslides, debris flows, and debris avalanches. Additionally, glaciers contribute substantially to water resources, which can be substantially reduced in many areas of the world [47].

6. Microorganisms in retreat glaciers

Global warming is having a great impact on glaciers, because of the change in air temperature and precipitation [48, 49]. Glacier retreat directly affects various atmospheric, climatic, and ecological phenomena. In retreat glaciers, the thickness of ice and snow decreases, and fossil ice emerges, forefield surface increases (**Figure 4**), new soil develops; and they are colonized by new prokaryotic and eukaryotic microbial species.

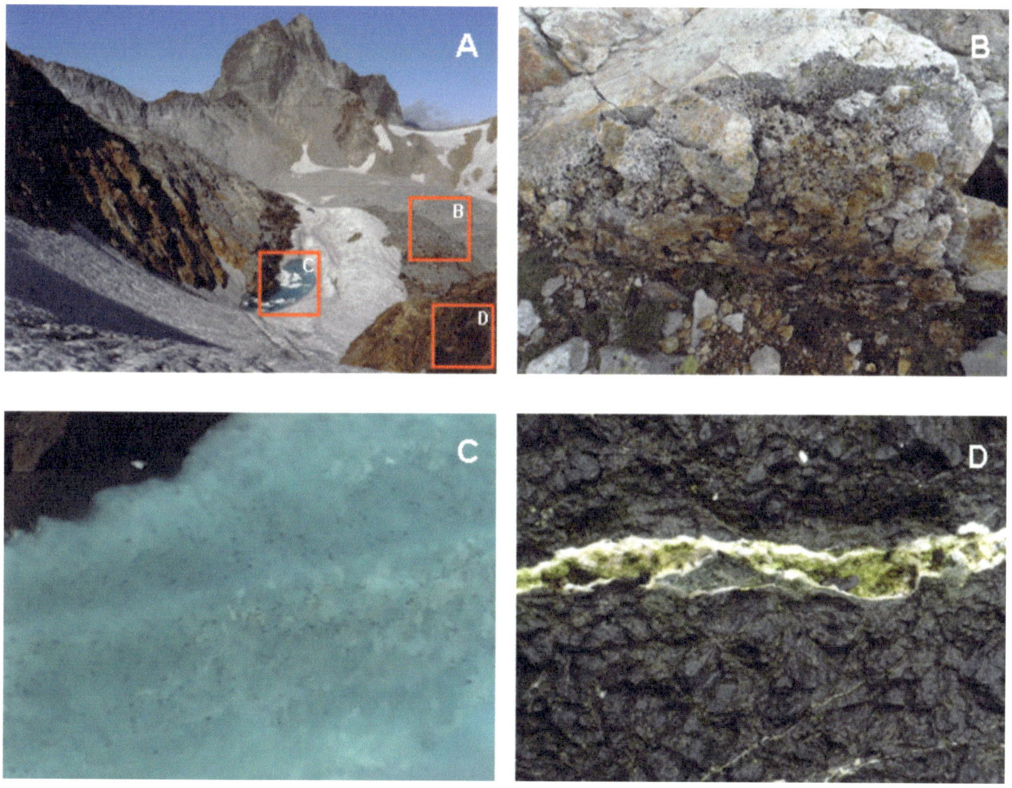

Figure 4. Retreat glacier. (A) In retreat glaciers, the thickness of ice and snow decreases, and it is colonized by new species. (B) Rock with lichens. (C) Microorganisms in the newly formed lake. (D) A rock showing the layer of endolithic phototrophic green algae.

The consequences of climate change are different according to the type of glacier. Depending on the location of the glaciers, these can be classified as terrestrial and marine (**Figure 2**). From the terrestrial glaciers, new lakes and rivers are shaped by runoff waters. On the contrary, some glaciers have a marine margin and terminate in a calving front. In glaciers ending on land, there is continuous permafrost at ice front, while calving glaciers present partly or completely temperate tidewater tongues [4].

One of the effects of climate change on glaciers is that the glacial ice melts and disappears, and microbial communities inhabiting them are being seriously affected [50]. Global warming is changing the basal temperature of the ice, going from cold to polythermal, which causes the growth of new microorganisms that are not psychrophiles but mesophiles thus leading to changes in the diversity and composition of microbial populations [51, 52].

The microorganisms that inhabit glaciers can also contribute to the production of heat as a consequence of their metabolic activity [53]. The amount of heat produced in cryoconite holes of glacial surface has been quantified and reaches 10% of the heat that melts the cryoconite walls during the summer [54]. Although these works have been much questioned [22], a recent work by Hollesen et al. [55] has shown that bioheat can accelerate ice melting on glacier surfaces.

Microorganisms play a main role in glaciers, mainly carrying out key processes in the development of soil, biogeochemical cycling and facilitating plant colonization when glaciers have ultimately retired.

6.1. Development of soil

Global climate change has accelerated glacial retreat. When the glacier ice melts and disappears, recently deglaciated soils establish a new ecosystem at the glacier forefield. Microorganisms are the initial colonizers of these recently exposed soils [7]. Thanks to their metabolic activity, new molecules are obtained that act as nutrients [7]. The microbial community of the newly formed soil is composed of heterotrophic microorganisms, autotrophic microorganisms, and nitrogen-fixing diazotrophs. Allochthonous material is derived from the glacier surface [21, 56], precipitation and aerial deposition [57] and biological sources such as mammal and bird droppings [58]. Additionally, adjacent ecosystems such as marine and subglacial environments are likely to contribute to the nutrient dynamics [58–60].

Downstream of the glacier, torrents are formed from the runoff water. These watercourses carry mineral salts and organic matter that allow the growth of new microorganisms. Biofilms grow on the banks of the streams, containing new microbial communities that although may remain psychrophilic, begin to have majority of psycrotrophic or mesophilic microorganisms.

Endolithic phototrophs, especially green algae and cyanobacteria, grow inside rocks, inhabiting porous rocks near the glacier surface [24]. Rocks are heated by the sun, and water from snow melt can be absorbed, supplying moisture needed for the growth of microorganisms. In addition to being free-living phototrophs, green algae and cyanobacteria coexist with fungi in endolithic lichen communities. Metabolism and growth of these internal rock communities slowly weathers the rock, allowing gaps to develop where water can enter, freeze, and thaw, and eventually crack the rock, producing new habitats for microbial colonization. The decomposing rock also forms a crude soil that can support the development of plant and animal communities in environments where conditions (temperature, moisture, and so on) allow [24].

The ice from the glaciers draws till, forming moraines around and inside the glacier. But it also draws organic matter from bird droppings and from dead plants and animals. In the development of soil, microorganisms break down this organic matter and produce carbon dioxide, water, and heat. Bacteria are responsible for a very little amount of the heat generation in ice, using a broad range of enzymes to chemically break down a variety of organic materials. Many bacteria are motile and can move into the ice channels of permafrost. When conditions become unfavorable, some bacteria survive by forming endospores, which are highly resistant to the cold and the lack of water and food sources. Microbial eukaryotes such as fungi are important because they break down debris, enabling bacteria to continue the decomposition process. They spread and grow by producing many cells and filaments. They can attack organic residues that are too dry or low in nitrogen for bacterial decomposition. Molds are strict aerobes that grow both as unseen filaments and as black, gray, or white fuzzy colonies on the surface. Most fungi are saprophytes; they live on dead material and obtain energy by breaking down organic matter. At last, protists obtain their food from organic matter in the same way as bacteria do but also act as secondary consumers ingesting bacteria and fungi.

6.2. Plant colonization

Retreating glacier fronts expose large expanses of deglaciated forefield, which become colonized by microbes and plants. The space that had been occupied by a glacier which only contained psychrophilic microorganisms is occupied primarily by mesophilic microorganisms inside and on the ice. When this ice disappears and soil begins to develop, rocks and tilt emerge in the moraines; this soil is colonized by lichens on rocks, by algae in streams and by higher plants and animals on the forefield. Most green algae inhabit freshwater, while others are found in moist soil [24]. Other green algae live as symbionts in lichens growing on rocks. In the newly formed soil, mainly consisting of permafrost, the growth of small plants begins. Their roots fragment the ground, forming small channels through which the water that carries ions in solution runs. In this way, the permafrost is fragmented, and it freezes less and less.

In glacier forelands, soil microorganisms are essential for plant growth as they play a key role in the nutrient cycling. In this phase, nitrogen, phosphorus, and other nutrients accumulate and facilitate succeeding plant growth [61]. Nitrogen-fixing plants are common in the primary succession of newly deglaciated soils [62]. Such plant-driven changes to soil nitrogen cycling have significant effects on the establishment of subsequent plant communities [63]. Rhizosphere microbial communities are fundamental for soil cycling, and they are mainly dominated by Proteobacteria, Bacteroidetes, Acidobacteria, Actinobacteria [64], and Firmicutes [65].

6.3. Biogeochemical cycling

Given their coverage at a global scale, snow and ice could have a major and underestimated role in global biogeochemical cycling [14]. It is essential to know how climate change is shaping the distribution and diversity of microbial communities, since microorganisms are very important components in several biogeochemical processes [66] and in food webs [67].

Nutrient matter in retreat glaciers is variable. The carbon content in forefields is very little. It comes from three distinct sources: autochthonous primary production by autotrophic microorganisms; the deposition of allochthonous material; and ancient organic pools derived from under the glacier [7]. Carbon dioxide is removed from the atmosphere primarily by photosynthesis of snow algae and cyanobacteria, and marine microorganisms in marine glaciers; and it is returned to the atmosphere by chemoorganotrophic microorganisms. Glaciers also provide organic matter to downstream ecosystems [7].

Other important nutrients in forefield soils such as nitrogen in the forms of nitrate, nitrite, and ammonia are microbially fixed from atmospheric nitrogen by cyanobacteria and some other bacteria. There are also external sources such as snowmelt, aerial deposition, and the breakdown of complex organic material or sedimentary bedrocks [7]. Bioavailable phosphorus and iron are usually abundant in the topsoil or bedrock of glaciated regions from weathering of the mineral surface [5].

6.4. Acidification or alkalinization of runoff waters

Another important effect of glacier retreat is the modification in the chemical composition of the runoff water of the glaciers. When a part of the terrain that had been covered by the glacier is open to the elements, rocks appear on the surface, whose minerals dissolve and change the physical and chemical characteristics of the runoff water. Occasionally, the waters become even toxic because of the presence of heavy metals [68]. Sometimes the formation of alkaline lakes can be observed, due to the mineral salts entrained by the water of runoff in the ground that had been occupied by a glacier, like the Amarga Lagoon in Chile (**Figure 5(A)**). These lagoons can be inhabited by cyanobacteria which grow forming laminar colonies whose calcareous skeletons fossilize generating sedimentary rocks named stromatolites (**Figure 5(B)**). They are formed when cells build up a carbonate skeleton, integrating particles present in the lake water.

Otherwise, when certain minerals such as pirite are exposed to air and water, a slow chemical reaction with molecular oxygen occurs. While this abiotic reaction can lead to the development of acidic conditions, the degree to which acid mineral drainage becomes an overwhelming burden on the environment results from the oxidative dissolution, a reaction catalyzed by microorganisms [69]. This process affects differently to the terrestrial and marine glaciers. In land glaciers, the runoff waters become more acidic, which affects the rivers and lakes that receive their waters. This can affect the flora and fauna, the crops and the human populations that live downstream. When this fact affects the marine glaciers, the composition of marine tidewater tongues changes their chemical composition: water salinity decreases, and at the same time, water becomes more acidic. Acidification of sea water impacts ocean species to varying degrees. A more acidic environment has a dramatic effect on some calcifying species, including oysters, clams, sea urchins, corals, and calcareous plankton [70].

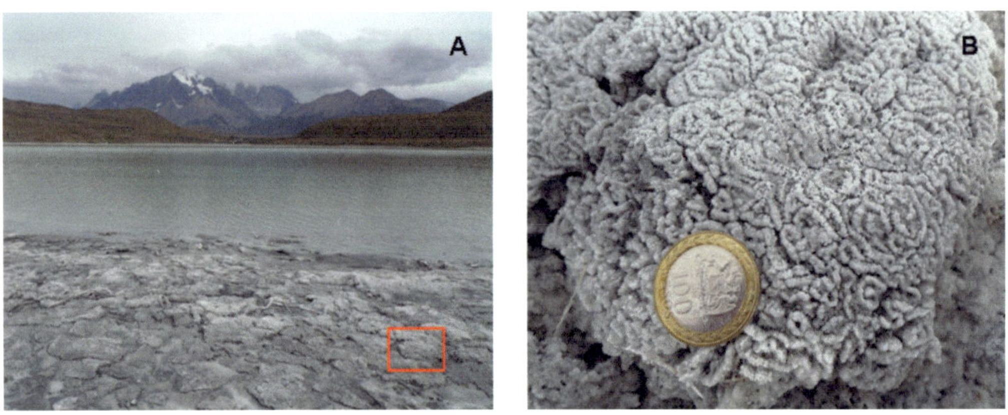

Figure 5. Alkalinization of runoff waters. (A) Formation of an alkaline lake (pH 9.1) due to the mineral salts entrained by the water of runoff in the ground that had been occupied by a glacier. The inset shows the location of **Figure 5(B)**. (B) Calcareous skeleton of a stromatolite of Amarga Lagoon at Chile.

7. Microorganisms as indicators of the conservation status of glaciers

In the last few decades, recently deglaciated areas present in different glacial zones in the world, are available for colonization and primary succession, especially initiated by pioneer microorganisms [71] followed by plants [72] and animals [73].

In retreat glaciers, the microbial populations of the glaciers are very different from those found in the surrounding soils. Some reports [61] have demonstrated that both bacterial and fungal community structures show significant differences between deglaciated sites and successional sites that had been ice-free over more than 100 years [74]. These changes have a strong influence on the processes of colonization and succession in the areas where glacier ice has melted [72]. Firstly, microbial communities change as psychrophilic microorganisms are replaced by mesophilic microorganisms. Then, plants and animals colonize these newly formed environments. Additionally, characteristic bacterial species can be found in each glacier zone and not found in the others. So, there are "type species" that can subsist thanks to their special metabolism and molecular mechanisms of adaptation. An example is shown in **Figure 6**, in which the population of microorganisms in three habitats: glacier snow [51, 75, 76], glacier front [6], and forefield [6, 62, 71] are compared. Although the compared glaciers are located in very different places around the world, it can be observed that the distribution of the main groups of microorganisms is different for each of the three habitats.

7.1. Glacier snow

Several reports have been published [77] about the little microbial abundance observed on glacier snow. For example, in Alpine snow packs, bacterial abundances range between 10^3 and10^5 cells/ml [78, 79], and in Svalbard archipelago, snow bacterial abundances are about 2×10^4 cells/ml [75, 80]. In cell counts performed on Antarctic snow, it was observed that the microbial abundance was even lower ($<10^3$ cells/ml).

Regarding microbial diversity in glacier snow (**Figure 6(A)**), the effect of snow melt on bacterial community structure and diversity of surface environments of a Svalbard glacier has been

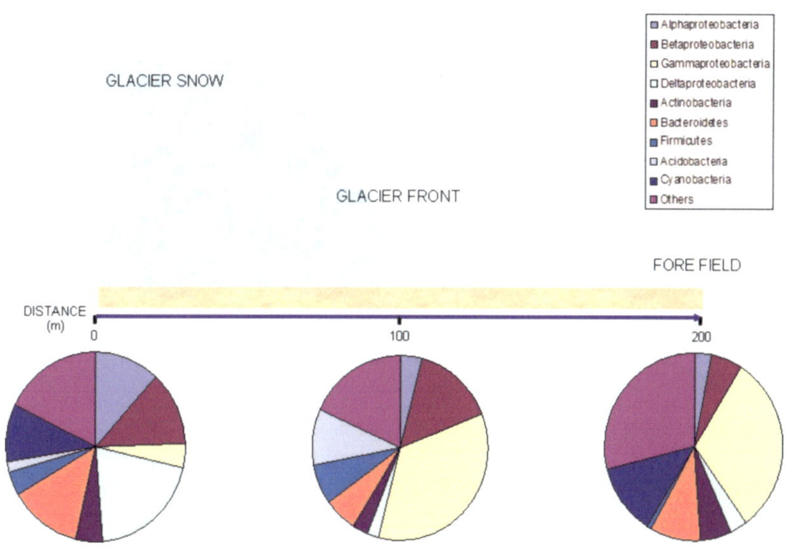

Figure 6. Bacterial community structure along a glacier front based on 16S rRNA gene sequences. Pie charts represent relative abundances of bacterial classes for three glacier environments: glacier snow, glacier front and forefield. The data come from Refs. [51, 76, 77] for glacier snow, from Ref. [6] for glacier front and from Refs. [6, 62, 74] for forefield.

examined using analyses of 16S rRNA genes [51]. In these studies, it was observed that the bacterial community structure depends on the type of snow deposition. However, the most interesting fact, from the point of view of monitoring the state of conservation of the glacier is that slush (the product of decomposition of snow when it melts) contains lineages of bacteria completely different from those of freshly fallen snow, which implies a change in the composition of the community structure that is post-depositional.

Other studies carried out in Greenland demonstrated that the phylogenetic composition of the microbial communities was different within the snow layers [75]. Proteobacteria, Bacteroidetes, and Cyanobacteria dominated in the middle and top snow layers, although Actinobacteria and Firmicutes were also abundant. In the deepest snow layer, large percentages of Firmicutes and Fusobacteria were found [75]. Large numbers of eukaryotic chloroplasts belonging to Streptophyta and Chlorophyta were also observed, demonstrating that microeukaryotes were also present in snow. Cyanobacteria and algae were almost exclusively found in the top and middle layers of the snow pack which are probably feeding the heterotrophic members of the microbial communities.

Some reports have demonstrated that the composition of snow microbial communities depends on the proximity to the sea [76]. In glacier snow, typical species of marine environments such as the Alphaproteobacteria have been found in samples from Antarctica, although Bacteroidetes and Cyanobacteria are also present [76].

7.2. Glacier front

Microbial communities in glacier fronts have been especially studied in the Antarctic Peninsula which is among the regions with the fastest warming rates, and where regional climate change has been linked to an increase in the mean rate of glacier retreat [6].

Archaeal and bacterial 16S rRNA gene sequences obtained from soil samples collected in the Wanda Glacier forefield showed that the diversity and richness were surprisingly high, and that communities were dominated by Proteobacteria, Bacteroidetes, and Euryarchaeota, with many archaeal and bacterial phylotypes yet unclassified (**Figure 6(B)**). Some of the phylotypes found were also related to marine microorganisms, indicating the importance of the marine environment as a source of colonizers for these recently deglaciated environments [6].

Concerning microbial abundance, some examples have been published. In Greenland glacier fronts, between 6 and 30 × 10^7 cells/ml, it has been reported [77].

7.3. Fore field

It has been published that microbial abundance in an Antarctic glacier (Ecology Glacier) forefield is increased along several sampling points from the glacier front to the farther outskirts of the glacier [71]. The same effect has been observed in the Peruvian Andes glaciers, where abundances of Cyanobacteria and Diatoms increased over the time of succession [62].

Regarding diversity, new soils from recently deglaciated soils are colonized by a diverse community of microorganisms during the first years following glacial retreat. Taxonomically microorganisms from Ecology Glacier forefield [71] belonged to the alpha, beta, and gamma subdivisions of the Proteobacteria and to the Cytophaga-Flavobacterium-Bacteroides (CFB) group (**Figure 6(C)**). Filamentous fungi were relatively abundant and represented mainly by oligotrophs.

In the recently deglaciated areas of the Peruvian Andes [62], it has been observed that a significant increase in cyanobacterial diversity corresponded with increases in soil stability, heterotrophic microbial biomass, soil enzyme activity, and the presence of photosynthetic and photoprotective pigments.

In glaciers, increasing temperature leads to a rapid retreat of ice, which increases water production [45, 72]. In glacier forefields, the runoff water of the glaciers can origin rivers and lakes [81]. For example, in the High-Arctic, it has been reported that Bacteroidetes, Actinobacteria, and Verrucomicrobia were the most abundant phyla in freshwater, while relatively few Proteobacteria and Cyanobacteria were present. Possibly, light intensity controlled the distribution of the Cyanobacteria and algae which in turn fed the heterotrophic bacteria [75].

Photosynthetic and nitrogen-fixing microorganisms play an important role in acquiring nutrients and facilitating ecological succession in soils during the first years of succession, many years before the establishment of mosses, lichens, or vascular plants [62]. Afterward, species of green soil algae are important pioneers in the colonization process of the areas recently denuded of ice [72].

At last, soil macrofauna and mesofauna colonize the fore fields. The successional chronosequence of an Alpine glacier was studied at several stages from 4 to 150 years of age since deglaciation [73]. Within the first 50 years, macrofauna biomass and mesofauna abundance increased rapidly, and successional age was the major determinant of community composition [73]. Some studies about soil mesofauna in high alpine ecosystems of the Central Alps demonstrated the shifts in species richness and density of arthropod such as oribatid mites [82]. In newly formed soils, some arthropods populate new soils, which in turn, promote the growth of fungi and bacteria and contribute to the formation of the new soil microstructure [82]. Nevertheless, these new fungi and bacteria are different from those that used to live in glaciers, as the novel species of plants and animals, contain associated microorganisms; for example, new microorganisms contained in animal droppings or symbiotic rhizosphere microbial communities associated to plants [65].

Microbial ecology can be a tool for monitoring the biological change that happens in retreat glaciers. Ecological researches conducted along deglaciated chronosequences in some glaciers have been carried out in order to understand the development of ecosystems. In these studies, distance from a receding glacier is used as a proxy for soil age, with older soils being further from the glacier front [62].

8. Conclusion

In summary, glaciers are retreating in many areas of the world due to global warming, and many of them will be severely affected or will disappear in a few years. Glaciers are unique biomes dominated by microbial communities which maintain active biochemical routes. Their metabolic activity plays an important role in glaciers, mainly carrying out key processes in the development of soil, changing biogeochemical cycling, altering the composition of runoff waters and facilitating plant and animal colonization when glaciers have ultimately retired. These processes impact the planet not only locally but also globally. Microorganisms are perfectly adapted to their harsh environment and are very susceptible to environmental changes. Colonization and primary succession of a recently deglaciated area implies that the abundance of microorganisms increases along deglaciated areas. Yet, at the same time, the diversity of microbial populations changes. In many cases, the number of different species may be lower than it is in the glacier. Thus, abundance and distribution of microorganisms can be considered indicative of the conservation status of glaciers, because alterations in their abundance and distribution depend on glacier conditions. Microbial ecology can be a tool for monitoring the biological change that happens in retreat glaciers.

Acknowledgements

We are indebted to Paula Alcazar for her technical assistance. This research was supported by grants CTM/2008-00304/ANT, CTM 2010-12134-E/ANT and CTM2011-16003-E from the Spanish Ministerio de Economía y Competitividad (MINECO). E.G.L. is recipient of a MINECO Fellowship (Grant PTAT2010-03424, Programa Nacional de Contratación e Incorporación de RRHH).

Author details

Eva Garcia-Lopez and Cristina Cid*

*Address all correspondence to: cidsc@inta.es

Microbial Evolution Laboratory, Center for Astrobiology (CSIC-INTA), Madrid, Spain

References

[1] Anesio AM, Laybourn-Parry J. Glaciers and ice sheets as a biome. Trends in Ecology & Evolution. 2012;**27**:219-225

[2] Hodson A, Anesio AM, Tranter M, Fountain A, Osborn M, Priscu J, Laybourn-Parry J, Satler B. Glacial ecosystems. Ecological Monographs. 2008;**78**:41-67

[3] Hagen JO, Liestøl O, Roland E, Jørgensen T. Glacier Atlas of Svalbard and Jan Mayen. Nor. Polarinst. Medd. 129. Oslo: Norwegian Polar Institute; 1993

[4] Hagen JO, Kohler J, Melvold K, Winther JG. Glaciers in Svalbard: Mass balance, runoffand freshwater flux. Polar Research. 2003;**22**:145-159

[5] Boetius A, Anesio AM, Deming JW, Mikucki JA, Rapp JZ. Microbial ecology of the cryosphere: sea ice and glacial habitats. Nature Reviews Microbiology. 2015;**13**:677-690

[6] Pessi IS, Osorio-Forero C, Galvez EJC, Simões FL, Simões JC, Junca H, Macedo AJ. Distinct composition signatures of archaeal and bacterial phylotypes in the Wanda Glacier forefield, Antarctic Peninsula. FEMS Microbiology Ecology. 2015;**91**:1-10

[7] Bradley JA, Singarayer JS, Anesio AM. Microbial community dynamics in the forefield of glaciers. Proceedings of the Royal Society B. 2014;**281**:20140882

[8] Schüte UM, Abdo Z, Bent SJ, Williams CJ, Schneider GM, Solheim B, Forney LJ. Bacterial succession in a glacier foreland of the High Arctic. The ISME Journal. 2009;**3**:1258-1268. DOI: 10.1038/ismej.2009.71

[9] Rodrigues DF, Tiedje JM. Coping with our cold planet. Applied and Environmental Microbiology. 2008;**74**:1677-1686

[10] Russell NJ. Cold adaptation of microorganisms. Philosophical Transactions of the Royal Society of London. 1990;**326**:595-611

[11] Doyle S, Dieser M, Broemsen E, Christner B. In: General Characteristics of Cold-Adapted Microorganisms. Polar Microbiology: Life in a Deep Freeze. Robert VM, Lyle G, editors. Washington, DC: Whyte ASM Press; 2012. pp. 103-125

[12] Marshall SJ. The Cryosphere. New Jersey: Princeton University Press; 2011

[13] GLIMS, National Snow and Ice Data Center. GLIMS Glacier Database. Boulder, Colorado, USA: National Snow and Ice Data; 2012

[14] Maccario L, Sanguino L, Vogel TM, Larose C. Snow and ice ecosystems: Not so extreme. Research in Microbiology. 2015;**166**:782-795

[15] Cavicchioli R, Siddiqui KS, Andrews D, Sowers KR. Low-temperature extremophiles and their applications. Current Opinion in Biotechnology. 2002;**13**:253-261

[16] D'Amico S, Collins T, Marx JC, Feller G, Gerday C. Psychrophilic microorganisms: Challenges for life. EMBO Reports. 2006;**7**:385-389

[17] Garcia-Descalzo L, Alcazar A, Baquero F, Cid C. Identification of in vivo HSP90interacting proteins reveals modularity of HSP90 complexes is dependent on the environment in psychrophilic bacteria. Cell Stress & Chaperones. 2011;**16**:203-218

[18] Alcazar A, Garcia-Descalzo L, Cid C. Microbial evolution and adaptation in icy worlds. In: Astrobiology: Physical Origin, Biological Evolution and Spatial Distribution. New York: Springer Verlag Inc; 2010. pp. 81-95

[19] Garcia-Descalzo L, Alcazar A, Baquero F, Cid C. Biotechnological Applications of Cold-Adapted Bacteria. Extremophiles: Sustainable Resources and Biotechnological Implications. New Jersey: Wiley-Blackwell ed. Hoboken; 2012. pp. 159-174

[20] Tranter M, Brown GH, Raiswell R, Sharp MJ, Gurnell AM. A conceptual model of solute acquisition by Alpine glacial meltwaters. Journal of Glaciology 1993;**39**:573-581

[21] Fountain AG, Nylen TH, Tranter M, Bagshaw E. Temporal variations in physical and chemical features of cryoconite holes on Canada Glacier, McMurdo Dry Valleys, Antarctica. Journal of Geophysical Research Biogeoscience. 2008;**113**:G01S92

[22] Cook J, Edwards A, Takeuchi N, Irvine-Fynn T. Cryoconite: The dark biological secret of the cryosphere. Progress in Physical Geography 2016;**40**:66-111

[23] Mieczen T, Gorniak D, Swiatecki A, Zdanowski M, Tarkowska-Kukuryk M, Adamczuk M. Vertical microzonation of ciliates in cryoconite holes in Ecology Glacier, King George Island. Polish Polar Research. 2013;**34**:201-212

[24] Madigan MT, Martinko JM, Stahl DA, Clark DP. Brock Biology of Microorganisms 13th ed. San Francisco, CA: Pearson Education, Inc; 2012

[25] Siefirmann-Harms D. The light-harvesting and protective functions of carotenoids in photosynthetic membranes. Physiologia Plantarum 1987;**69**:501-568

[26] Martin-Cerezo ML, Garcia-Lopez E, Cid C. Isolation and identification of a red pigment from the Antarctic bacterium Shewanella frigidimarina. Protein and Peptide Leters 2015;**22**:1076-1082

[27] Becker-Hapak, M, Troxtel, E, Hoerter, J, Eisenstark, A. RpoS dependent overexpression of carotenoids from *Erwinia herbicola* in OXYR-deficient *Escherichia coli*. Biochemical and Biophysical Research Communications. 1997;**239**:305-309

[28] Goodwin TW. The Biochemistry of Carotenoids. London: Chapman and Hall; 1980

[29] Lut S, Anesio AM, Villar J, Benning LG.. Variations of algal communities cause darkening of a Greenland glacier. FEMS Microbiology Ecology. 2014;**89**:402-414

[30] Jagannadham MV, Chatopadhyay MK, Subbalakshmi C, Vairamani M, Narayanan K, Rao CM, Shivaji S. Carotenoids of an Antarctic psychrotolerant bacterium, *Sphingo bacterium antarcticus*, and a mesophilic bacterium, *Sphingobacterium multivorum*. Archives of Microbiology. 2000;**173**:418-424

[31] Vollmers J, Voget S, Dietrich S, Gollnow K, Smits M, Meyer K, BrinkhoffT, Simon M, Daniel R. Poles apart: Arctic and Antarctic *Octadecabacter* strains share high genome plasticity and a new type of xanthorhodopsin. PLoS One. 2013;**8**:e63422

[32] Garcia-Descalzo L, Garcia-Lopez E, Alcazar A, Baquero F, Cid C. Proteomic analysis of the adaptation to warming in the Antarctic bacteria Shewanella frigidimarina. BBA Proteins and Proteomics. 2014;**1844**:2229-2240

[33] Garcia-Lopez E, Cid C. Color producing extremophiles. In: Bio-Pigmentation and Biotechnological Implementations. ISBN: 978-1-119-16614-6. New Jersey: Wiley-Blackwell ed. Hoboken; 2016

[34] Gunde-Cimerman N, Sonjak S, Zalar P, Frisvad JC, Diderichsen B, Plemenitas A. Extremophilic fungi in arctic ice: A relationship between adaptation to low temperature and water activity. Physics and Chemistry of the Earth. 2003;**28**:1273-1278

[35] Price PB. Microbial life in glacial ice and implications for a cold origin of life. FEMS Microbiology Ecology. 2007;**59**:217-231

[36] Price PB. A habitat for psychrophiles in deep, Antarctic ice. PNAS. 2000;**97**:1247-1251

[37] Tung HC, Bramall NE, Price BP. Microbial origin of excess methane in glacial ice and implications for life on Mars. PNAS. 2005;**102**:18292-18296

[38] Butinar L, Spencer-Martins I, Gunde-Cimerman N.Yeasts in high Arctic glaciers: The discovery of a new habitat for eukaryotic microorganisms. Antony van Leeuwenhoek. 2007;**91**:277-289

[39] Lerner L, Wilmoth B. Chemoautotrophic and chemolithotrophic bacteria. In: World of Microbiology and Immunology. Lerner K, editors. 2016. Available from: htp://www.encyclopedia.com [Accessed: 2016-12-12]

[40] Leuliete E, Miler L. Closing the sea level rise budget with altimetry, Argo, and GRACE, Geophysical Research Leters. 2009;**36**:L04608

[41] Kwok R, Cunningham GF. Variability of Arctic sea ice thickness and volume from CryoSat-2. Philosophical Transactions of the Royal Society A. 2015;**373**:20140157. DOI: 10.1098/rsta.2014.0157

[42] Holland PR, Kwok R. Wind-driven trends in Antarctic sea ice drift. Nature Geoscience. 2012;**5**. DOI: 10.1038/NGEO1627

[43] Kwok R, Untersteiner N. The thinning of Arctic sea ice. Physics Today. 2011;**64**:36-41

[44] Kwok R, Rothrock DA. Decline in Arctic sea ice thickness from submarine and ICESat records: 1958-2008. Geophysical Research Leters. 2009;**36**:L15501. DOI: 10.1029/2009GL039035

[45] Rignot E, Bamber JL, van den Broeke MR, Davis C, Li Y, van de Berg WJ, van Meijgaard E. Recent Antarctic ice mass loss from radar interferometry and regional climate modeling, Natural Geoscience. 2008;**1**:106-110

[46] Velicogna I. Increasing rates of ice mass loss from the Greenland and Antarctic ice sheets revealed by GRACE. Geophysical Research Leters. 2009;**36**:L19503

[47] Immerzeel WW, van Beek LPH, Bierkens MFP. Climate change will affect the Asian water towers. Science. 2010;**328**:1382. DOI: 10.1126/science.1183188

[48] Førland EJ, Benestad F, Flatøy F, Hanssen-Bauer I, Haugen JE, Isaksen K, Sorteberg A, Ådlandsvik B. Climate development in north Norway and the Svalbard region during 1900-2100. Norwegian Polar Institute Report Series 128. Tromsø: Norwegian Polar Institute; 2009

[49] Solomon S, Qin D, Manning M, Chen M, Marquis KB, Averyt, Tignor M, Miller HL. The physical science basis: contribution of Working Group I to the fourth assessment report of the Intergovernmental Panel on Climate Change. Climate change 2007 (ed). Cambridge: Cambridge University Press; 2007

[50] Zeng YX, Yan M, Yu Y, Li HR, He JF, Sun K, Zhang F. Diversity of bacteria in surface ice of Austre Lovénbreen glacier, Svalbard. Archives Microbiology. 2013;**195**:313-322

[51] Hell K, Edwards A, Zarsky J, Podmirseg SM, Girdwood S, Pachebat JA, Insam H, Satler B. The dynamic bacterial communities of a melting High Arctic glacier snowpack. The ISME Journal. 2013;**7**:1814-1826

[52] Nowak A, Hodson A. Changes in meltwater chemistry over a 20-year period following a termal regime switch from polythermal to cold-based glaciation at Austre Broggerbreen, Svalbard. Polar Research. 2014;**33**:22779

[53] Gerdel RW, Drouet F. The cryoconite of the Thule area, Greenland. Transactions of the American Microbiology Society. 1960;**79**:256-272

[54] McIntyre NF. Cryoconite hole thermodynamics. Canadian Journal of Earth Science. 1984;**21**:152-156

[55] Hollesen J, Mathiesen H, Møller AB, Elberling B. Permafrost thawing in organic Arctic soils accelerated by ground heat production. National Climate Change. 2015;**5**:574-578. DOI: 10.1038/nclimate2590

[56] Anesio AM, Hodson AJ, Frit A, Psenner R, Satler B. High microbial activity on glaciers: Importance to the global carbon cycle. Global Change Biology. 2009;**15**:955-960

[57] Larose C, Dommergue A, Vogel TM. Microbial nitrogen cycling in Arctic snowpacks. Environmental Research Leters. 2013;**8**:035004

[58] Mindl B, Anesio AM, Meirer K, Hodson AJ, Laybourn-Parry J, Sommaruga R, Satler B. Factors influencing bacterial dynamics along a transect from supraglacial runoffto proglacial lakes of a high Arctic glacier. FEMS Microbiology Ecology. 2007;**59**:307-317

[59] Kastovska K, Elster J, Stibal M, Santruckova H. Microbial assemblages in soil microbial succession after glacial retreat in Svalbard (high Arctic). Microbial Ecology. 2005;**50**:396-407

[60] Skidmore ML, Foght JM, Sharp MJ. Microbial life beneath a high Arctic glacier. Applied and Environmental Microbiology. 2000;**66**:3214-3220

[61] Deiglmayr K, Philippot L., Tscherko D, Kandeler E. Microbial succession of nitratereducing bacteria in the rhizosphere of *Poa alpina* across a glacier foreland in the Central Alps. Environmental Microbiology. 2006;**8**:1600-1612

[62] Schmidt SK, Reed SC, Nemergut DR, Grandy AS, Cleveland CC, Weintraub MN, Hill AW, Costello EK, Meyer AF, NeffJC, Martin AM. The earliest stages of ecosystem succession in high-elevation (5000 metres above sea level), recently deglaciated soils. Proceedings of the Royal Society B. 2008;**275**:2793-2802

[63] Chapin FS, Walker LR, Fastie CL, Sharman LC. Mechanisms of primary succession following deglaciation at Glacier Bay, Alaska. Ecological Monographs. 1994;**64**:149-175

[64] Roesch LFW, Fulthorpe RR, Pereira AB, et al. Soil bacterial community abundance and diversity in ice-free areas of Keller Peninsula, Antarctica. Applied Soil Ecology. 2012;**61**:7-15

[65] Teixeira LC, Peixoto RS, Cury JC, Rosado AS. Bacterial diversity in rhizosphere soil from Antarctic vascular plants of Admiralty Bay, maritime Antarctica. The ISME Journal. 2010;**4**:989-1001

[66] Hoham RW, Duval B. Microbial ecology of snow and freshwater ice with emphasis on snow algae. In: Jones HG, Pomeroy JW, Walker DA, Hoham RW, editors. Snow Ecology: An Interdisciplinary Examination of Snow-Covered Ecosystems. UK: Cambridge University Press; 2001. pp. 168-228

[67] Garcia-Descalzo L, Garcia-Lopez E, Postigo M, Baquero F, Alcazar A, Cid C. Eukariotic microorganisms in cold enviroments: Examples from Pyrenean glaciers. Frontiers in Microbiology. 2013;**4**:55. DOI: 10.3389/fmicb.2013.00055

[68] Johnson DB, Hallberg KB. The microbiology of acidic mine waters. Research in Microbiology 2003;**154**:466-473

[69] Nordstrom DK. Mine waters: Acidic to circumneutral. Elements. 2011;**7**:393-398

[70] Pacific Marine Environmental Laboratory Carbon Program. Available from: htp://www.pmel.noaa.gov [Accessed: 30-01-2017]

[71] Grzesiak J, Żmuda-Baranowska M, Borsuk P, Zdanowski M. Microbial community at the front of Ecology Glacier (King George Island, Antarctica): Initial observations. Polish Polar Research. 2009;**30**:37-47

[72] Massalski A, Mrozińska T, Olech M. Ultrastructural observations on five pioneer soil algae from ice denuded areas (King George Island, West Antarctica). Polar Bioscience. 2001;**14**:61-70

[73] Kaufmann R, Fuchs M, Gosterxeier N. The soil fauna of an Alpine Glacier Foreland: Colonization and succession. Arctic, Antarctic, and Alpine Research. 2002;**34**:242-250

[74] Kim M, Jung JY, Laffly D, Kwon HY, Lee YK. Shifts in bacterial community structure during succession in a glacier foreland of the High Arctic. FEMS Microbiology Ecology. 2017;**93**. pii: fiw213

[75] Møller AK, Søborg DA, Al-Soud WA, Sørensen SJ, Kroer N. Bacterial community structure in high-arctic snow and freshwater as revealed by pyrosequencing of 16S rRNA genes and cultivation. Polar Research. 2013;**32**:17390

[76] Michaud L, Lo Giudice A, Mysara M, Monsieurs P, Raffa C, Leys N, Amalfitano S, Van Houdt R. Snow surface microbiome on the high antarctic plateau (DOME C). PLoS One. 2014;**9**:e104505. DOI: 10.1371/journal.pone.0104505

[77] Stibal M, Gözdereliler E, Cameron KA, Box JE, Stevens IT, Gokul JK, Schostag M, Zarsky JD, Edwards A, Irvine-Fynn TD, Jacobsen CS. Microbial abundance in surface ice on the Greenland Ice Sheet. Frontiers in Microbiology. 2015;**6**:225

[78] Alfreider A, Pernthaler J, Amann R, Sattler B, Glockner FO, Wille A, Psenner R. Community analysis of the bacterial assemblages in the winter cover and pelagic layers of a high mountain lake by in situ hybridization. Applied and Environmental Microbiology. 1996;**62**:2138-2144

[79] Sattler B, Puxbaum H, Psenner R. Bacterial growth in supercooled cloud droplets. Geophysical and Research Letters. 2001;**28**:239-242

[80] Amato P, Hennebelle R, Magand O, Sancelme M, Delort AM, Barbante C, Boutron C, Ferrari C. Bacterial characterization of the snow cover at Spitzberg, Svalbard. FEMS Microbiology Ecology. 2007;**59**:255-264

[81] Wilhelm L, Singer GA, Fasching C, Battin TJ, Besemer K. Microbial biodiversity in glacier- fed streams. The ISME Journal. 2013;**7**:1651-1660

[82] Fischer BM, Schatz H. Biodiversity of oribatid mites (Acari: Oribatida) along an altitudinal gradient in the Central Alps. Zootaxa. 2013;**3626**:429-454

Chapter 6

Glacier Forelands – Unique Field Laboratories for the Study of Primary Succession of Plants

Thomas Fickert

Additional information is available at the end of the chapter

http://dx.doi.org/10.5772/intechopen.69479

Abstract

While receding glaciers in several respects are related to negative consequences for resident people down river (e.g., water security, glacial lake outburst floods, etc.), from an ecological point of view, they are unique field laboratories to study vegetation development and dynamics from the very beginning. As the bare ground exposed by the receding glaciers generally does not provide a seed bank, the colonization of formerly ice-covered ground represents a true primary succession. There are two different approaches to study vegetation dynamics in glacier forelands: permanent plots and chronosequences ("space for time substitution"). As both procedures have their respective pros and cons, they should not be regarded to be mutually exclusive; rather, they should complement each other for a comprehensive understanding of the vegetation dynamics in glacier forelands. This chapter gives a general overview on patterns and processes of vegetation development in glacier forelands based on vegetation sampling in two glacier forelands of the Eastern Alps employing both abovementioned approaches. Sampling records groundcover of all vascular plants, structural features such as life form composition and dispersal biology types of the species as well as site characteristics. The two different approaches give evidence for both the early vegetation dynamics after deglaciation by the permanent plot studies and for potential future developments by the chronosequences.

Keywords: receding glaciers, vegetation dynamics, plant colonization, permanent plots, chronosequences

1. Introduction

The renowned German ecologist Heinz Ellenberg made a good point when stating "Nowhere can succession be studied more profitably than in the valley below the front of a large glacier" [1]. The opportunity to directly observe the vegetation development on new, hitherto unvegetated ground has fascinated botanists ever since, and in the European Alps, studies on vegetation development in glacier forelands date back well into the mid-nineteenth century. There are two fundamentally different approaches to study vegetation dynamics in glacier forelands: permanent plots and chronosequences ("space for time substitution" sensu [2], see **Figure 1**). Due to time constraints, the latter method is commonly employed, using spatially different sites to reconstruct a temporal sequence. In glacier forelands, dateable traces of the earlier extent of glaciers are commonly used [3], sometimes combined with lichenometric dating [4]. While the chronosequence approach is suitable to document shifts in species composition and vegetation structure as response to the time since melt-out, it does not give evidence how the colonization of bare ground is actually taking place. In addition, as different sites may be influenced by varying site histories, by unsimilar effects of the surrounding and/or topography (exposure, slope angle, etc.) or by differences in the frequencies and/or magnitudes of disturbances, not only the time since melt-out might be essential for the vegetation development observed.

A very accurate appraisal of the colonization dynamics in glacier forelands can be obtained by permanent plots. If an adequate number of resurveys is provided, permanent plot studies give good evidence on migration patterns, shifts in frequency or abundance of species, growth performance, and temporary setbacks. In addition, they allow to identify whether the development of species numbers or ground cover follows a more linear or more logarithmic trend and whether a trend reversal due to inter- and intraspecific competition at any point during succession occurs (see **Figure 2**). The downside of permanent plots is that a high level of patience is needed, a rare virtue in this day and age. Thus, it is no surprise that only few long-running permanent plot studies on primary succession in glacier forelands of the Alps (and beyond) exist.

As both procedures have their respective pros and cons, they should not be regarded to be mutually exclusive; rather, they should complement each other for a comprehensive treatment of the vegetation dynamics in glacier forelands. While the permanent plot studies give evidence for the early colonization by plants of the bare ground exposed by the receding glaciers, the chronosequences give hints for the long-term vegetation dynamics within glacier forelands. The own results presented here will be put into a larger context to give a general overview on patterns and processes of vegetation development in glacier forelands of the Alps. The following topics are of particular interest:

- How fast is the colonization of the bare ground taking place? Several studies showed that high-elevation plant species despite favorable dispersal modes (predominantly lightweight wind-dispersed seeds) show delayed vegetation dynamics due to high mortality rates during establishment (e.g., Refs. [5, 6]).

- Does colonization of the bare ground provided by receding glaciers follow a more linear or a more logarithmic trend (positive, i.e., delayed at the beginning, accelerated later on; negative, i.e., vice versa; see **Figure 2**)?

- Is there a point of trend reversal (e.g., decreasing species and/or individual numbers) due to increasing interspecific and intraspecific competition during succession (**Figure 2**)?

- How do site conditions (elevation, exposure, substrate, snow cover duration, microclimatic conditions, etc.) affect vegetation dynamics?
- Do facilitating or inhibiting interactions between species exist, and if so, at what moment do these become apparent [7–11]?

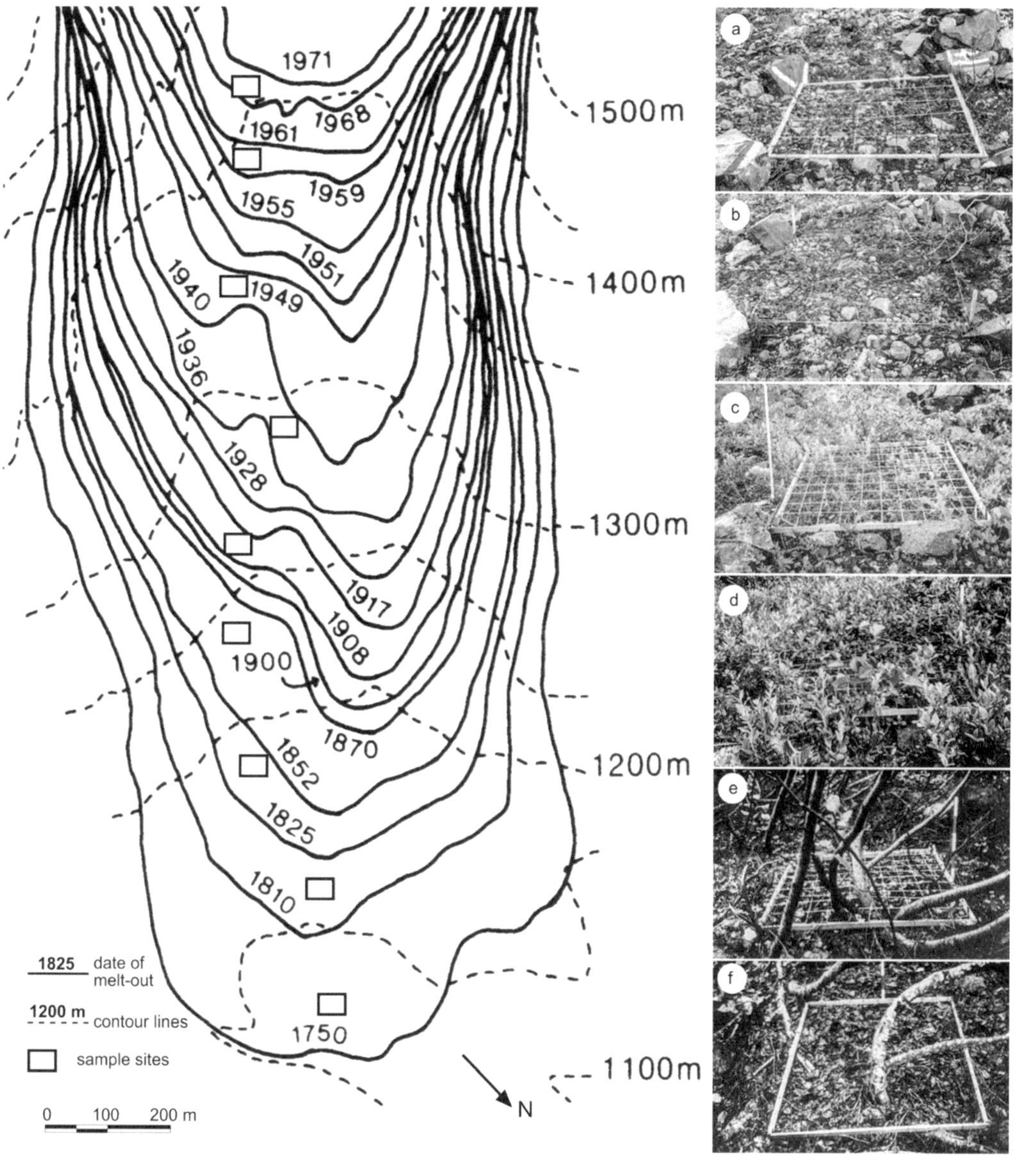

Figure 1. Juxtaposition of the two general procedures for studying succession of plants in glacier forelands: chronosequences using spatially different sites to reconstruct a temporal sequence (left, shown is Storbreen glacier in Norway) and permanent plots observing vegetation development on one and the same sample site (right, shown is Coopers Quadrat #1 at Grand Pacific Glacier, Alaska; a = 1921, b = 1935, c = 1949, d = 1955, e = 1967, f = 1982) (modified from Refs. [3, 17]).

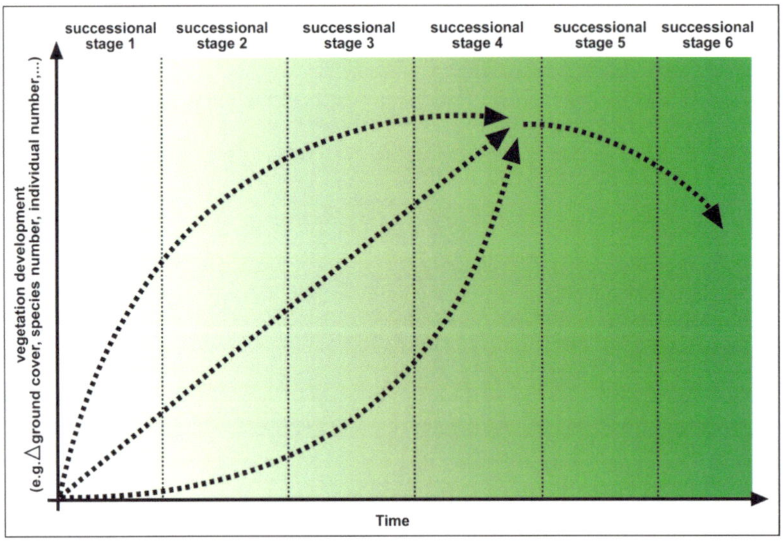

Figure 2. Hypothetical successional trajectories carrying linear or both positive and negative logarithmic/exponential characters. At a particular stage also reverse trends of decreasing species or individual numbers might occur.

2. Study sites, sampling design, and analyses

The two glacier forelands studied are located in the Central Eastern Alps. In summer 2005, permanent plots were established immediately in front of Goldbergkees (Hohe Tauern, Austria) and Lenksteinferner (Rieserferner, Italy), which were revisited every second year thereafter. In 2010, chronosequences were established within the same glacier forelands extending from the permanent plots down to moraines of the Little Ice Age (LIA) maximum which occurred around 1850. **Table 1** summarizes basic information on both study areas.

Sample sites are 10 m^2 (2 × 5 m) for both the permanent plot and the chronosequence studies and represent "mean" site conditions (i.e., no hollows with above-average snow cover duration or wind-exposed knolls with drier conditions). The permanent plots are arranged in two different sets. The A-sites were installed close to the glacier terminus (2–8 m in front of the ice margin in 2005), the B-sites 7–12 m away from the A-sites. Most sites became deglaciated in the year of the initial survey, and some of the B-sites probably already the year before. In total, 26 sample sites at Goldbergkees (13 A, 13 B) and 22 at Lenksteinferner (11 A, 11 B) are studied. Sites are GPS-recorded, flagged, and photo-documented for a precise match at resurveys. Vegetation sampling records groundcover and individual numbers of all vascular plants as well as structural features such as life form composition [12] and dispersal biology types [13] of the species. Mosses are sampled as undifferentiated species group. Taxonomy of vascular plant species follows [14]. Sampling occurs m^2-wise with the smallest unit being 0.01% ground cover (i.e., 1 cm × 1 cm on a 1 m^2 subplot). Raw data are subsequently converted to mean ground cover values as well as total number of species and individuals per 10 m^2 sample site.

	Goldbergkees	Lenksteinferner
Geology	Granitoid rocks	Granitoid rocks
Number of permanent plot samples	13 A, 13 B	11 A, 13 B
Elevation permanent plots	2390–2420 m a.s.l.	2600–2630 m a.s.l.
Elevation LIA terminal moraine	2180 m a.s.l.	2340 m a.s.l.
Number of chronosequence stages (three samples each)	8	9
Horizontal extent chronosequence	1300 m	1250 m
Vertical extent chronosequence	200 m	280 m

Table 1. Basic data on the permanent plot and chronosequence studies within the glacier forelands of Goldbergkees and Lenksteinferner.

The chronosequences extend from the permanent plots down to the LIA terminal moraines. For both glacier forelands detailed chronologies about glacier retreat since LIA maximum exists [15, 16], allowing for a quite accurate age determination of sample sites. At Goldbergkees, the total horizontal distance between the youngest and oldest sites is 1.3 km and 200 m elevational difference, on Lenksteinferner 1.25 km and 280 m, respectively. On Goldbergkees, eight different stages of time since melt-out were studied (2 years, 4 years, 15 years, 25–30 years, 55 years, 85 years, 120 years, 155 years) and on Lenksteinferner, nine different stages (2 years, 4 years, 20 years, 35 years, 55 years, 75 years, 90 years, 120 years, 155 years). Each stage is represented by three 10 m^2 plots. For the youngest stages (<10 year-ice-free), three of the permanent plot samples were used (three A-sites of 2007, three B-sites of 2009). Vegetation sampling is primarily the same as on the permanent plots, with the only modification that individual numbers are not counted. Environmental variables collected in situ for all sample sites of both approaches are elevation (by altimeter), exposure (by compass), slope angle (by clinometer), and rockiness of the ground (by visual estimation of coarse rocks > 6 cm in %).

For data analyses, univariate and multivariate statistical procedures were employed. To assess the successional development in glacier forelands quantitatively, documented changes between different samples are crucial—temporally different in the case of the permanent plots and spatially different in the case of the chronosequences. Primary succession in glacier forelands commonly starts with simple agglomerations of plants and subsequently becomes more and more complex. The increasing complexity during succession becomes obvious in change measures such as mean ground cover values (of singular species and total) as well as species and individual numbers. Temporal trends are derived by linear and/or nonlinear regressions. The range of variation in data sets (species numbers, individual numbers, groundcover, etc.) is depicted by box and whisker plots.

For detecting gradual changes in species composition within large data sets, ordination procedures assuming underlying gradients within data sets are appropriate tools. By means of similarity relationships, gradual changes of samples concerning species composition are calculated in a multidimensional ordination space. The aim of ordinations is to reduce the number of dimensions and to make complex datasets with many species and samples interpretable. For the chronosequence data, unconstrained linear principal component analyses (PCAs) are employed. The graphic presentation of ordination analyses is by two-dimensional scatter plots displaying samples and/or species; explanatory environmental variables (if available) are displayed as arrows. The arrows point from the origin of ordinates in the direction of samples with above average values of the particular variable; the length represents the relevance of the variable. Gradient analyses were performed with Canoco 4.5.

3. Early plant colonization in glacier forelands as revealed by permanent plot studies

After glacier retreat, new ground is provided for plant colonization. Substrate of the sample sites is variably rocky with amounts of rocks >6 cm between less than 25% to well over 80%. **Figure 3** portrays the early vegetation development in the glacier forelands of Goldbergkees and Lenksteinferner, showing basically similar trends concerning species numbers, individual numbers, ground cover, and life form composition.

At the initial survey in 2005, the A-sites immediately in front of the ice margin were entirely free of plants in both glacier forelands, while on the B-sites, a couple of species with very few individuals were already present. The early colonizers are the chamaephytes *Arabis alpina*, *Cerastium uniflorum*, *Saxifraga bryoides*, and *Saxifraga oppositifolia*; the herbs *Oxyria digyna* and *Veronica alpina*; as well as the grass *Poa alpina* on Goldbergkees and *A. alpina*, *Cerastium cerastoides*, *C. uniflorum*, the stoloniferous herb *Geum reptans*, and the grass *Poa laxa* on Lenksteinferner (**Figure 4**; for complete species lists of the two permanent plot studies, see Ref. [17]). Two years later also on the A-sites, the first individuals appear. On B-sites species and individual numbers exponentially increased, while ground cover is still low (**Figure 3**). The increasing trend continues during the first decade, slowing down slightly for species and individual numbers but accelerating for ground cover (**Figure 3**). In 2015, a total of more than 30 species are present on the permanent plots on Goldbergkees (median of individual A- and B-sites is 13 and 15, respectively) with over 1000 and 2000 individuals on the A- and B-sites, respectively. Somewhat lower are the values for the sample sites on Lenksteinferner with 22 species on the A-sites and 30 species on the B-sites (median of individual A- and B-sites is 8 and 11, respectively), with about half of the individual numbers of Goldbergkees (>500 on A-sites, just under 900 on B-sites).

Besides differences in absolute values concerning species numbers, individual numbers, and ground coverage (**Figure 3**), the permanent plot studies in the glacier forelands of Goldbergkees und Lenksteinferner prove a swift colonization of the bare ground and reveal generally similar trends in vegetation development. Vegetation dynamics in glacier forelands are controlled by three fundamental steps; all of them "may have, perhaps, the power to be the important limiting factor for succession and ecosystem development" [18]: step 1 is reaching the bare ground, step 2 is a successful establishment, and step 3 is growth and spreading.

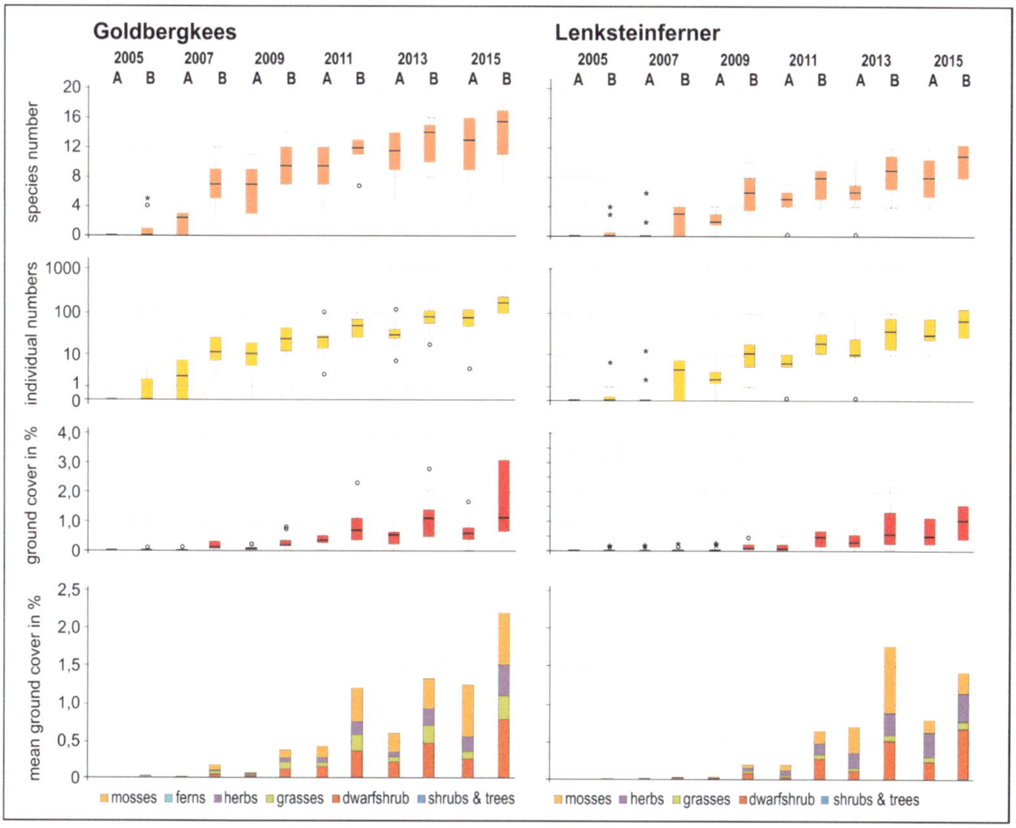

Figure 3. Vegetation dynamics on permanent plots (A- and B-sites) in the glacier forelands of Goldbergkees (left) and Lenksteinferner (right) between 2005 and 2015. Development and variation between sites are shown by box and whisker plots for species numbers, individual numbers, and ground cover (for vascular plants only); at the bottom bar plots show mean life form spectra.

3.1. Step 1 in primary succession: getting there

Reaching the bare ground is the first obstacle organisms have to conquer when colonizing new terrain. For this task, both a diaspore supply in the surroundings and the ability of a species to disperse from the seed source to the new ground are crucial. Concerning the former aspect, it makes a difference whether a glacier terminates within closed (sub)alpine vegetation with a rich diaspore supply or within sparsely vegetated scree slopes of the subnival belt. Several studies [19–22] have shown that recently deglaciated glacier forelands within the subalpine and lower alpine belt are colonized more quickly and more diverse concerning species numbers and life form composition than smaller glaciers terminating within the species poor upper alpine and subnival belts [1]. In addition, the few species present at higher elevations of the Alps do—despite small diaspores—not always possess high dispersal ability [5, 11], thus further impeding colonization. Seeding experiments, at least, have shown that artificial seed supply enhanced vegetation dynamics in glacier forelands and—what is even more important—that plant species are able to establish, which would hardly reach the bare

Figure 4. Characteristic early colonizers of central Alpine glacier forelands: *Arabis alpina* (a), *Cerastium uniflorum* (b), *Oxyria digyna* (c), *Saxifraga bryoides* (d), *Saxifraga oppositifolia* (e), stoloniferous *Geum reptans* (f), and *Poa alpina* (g) in the bulbil-producing ("viviparous") form.

ground by natural dispersal vectors [5, 11]. Nevertheless, commonly it does not take long until the first plant species in recently deglaciated high alpine glacier forelands (i.e., >2200 m a.s.l.) appear. Time frames reported range between 1 and 8 years [1, 21–24], and also the own survey attests a swift colonization with the first individuals appearing within 1–2 years [17]. The first colonizers are anemochorous taxa throughout, carried to the glacier foreland by valley winds from the surroundings and from lower elevations, too [25–27]. Thus, early colonizers are not only pioneer species sensu stricto (i.e., early colonizers not able to persist over time during succession, e.g., *S. oppositifolia*, *Saxifraga exarata*) but also early- to late-successional taxa (i.e., those that persist over time during succession but are able for early colonization as well, e.g., *A. alpina, C. uniflorum, O. digyna, S. bryoides*) and even those with ubiquitous behavior (e.g., *Leucanthemopsis alpina, Agrostis rupestris, P. alpina;* see Ref. [28]).

Chances for wind-dispersed early colonizers to reach the recently deglaciated glacier forelands are closely linked to the distance to seed sources. Even if taxa can be transported by wind over distances of up to a few kilometers (see Ref. [27]), highest seed rain is within a radius of several meters around the seed source [25, 27]. The longer the transport, the lower the chance to reach the new terrain, as the seeds may ground somewhere else or the diaspores are completely lost when carried to unsuitable sites for germination (water bodies, rocks, snow, dense established vegetation, etc.). The differences in species numbers, individual numbers, and ground cover values between the two study areas can be partly explained by differences in diaspore supply. Lenksteinferner is exposed to the North at a relatively high elevation (>2600 m a.s.l.) and surrounded by sparsely vegetated scree slopes with poor diaspore supply. More abundant seed sources are to be found at lower elevations from where seeds have to be carried up by valley winds. For instance, in 2015, a small larch seedling (*Larix decidua*) was encountered 300 vertical meters and more than 1 km away from the last cone-bearing adult Larch tree. Whether this seedling will survive remains to be seen; still, it highlights the relevance of long-distance dispersal for early colonization in glacier forelands. That Lenksteinferner is lagging behind Goldbergkees concerning species and individual numbers as well as groundcover values (**Figure 3**) most likely results from the larger distance to potential seed sources and thus the higher risk of diaspores getting lost. At Goldbergkees, located on lower elevation and in a more favorable exposure, last patches of closed alpine vegetation in the surrounding serve as seed sources for the colonization of the glacier foreland and are responsible for the higher species and individual numbers there. Besides wind, also water, avalanches, or mudflows might be locally important dispersal vectors for seeds or even whole plants into glacier forelands [26]. In addition, primary succession in glacier forelands might also be affected by seeds and plants originating from (debris-covered) glacier surfaces that are deposited in the glacier foreland after ice-melt [29–31].

3.2. Step 2 in primary succession: establishing

The second important step of primary succession in glacier forelands is a successful establishment of plants. This is not guaranteed for all diaspores that reach the glacier foreland, and there are high interspecific differences in seedling recruitment and survival [26, 32–34]. A study [32] has shown that in some species chances for germination are highest within the first year but declining shortly thereafter. For *Artemisia genipi* and *Achillea moschata*, germination success within the first year was at 98.8% and 68.8%, respectively. For other species, e.g., *Linaria alpina* or *S. oppositifolia*, germination success within the first year was very low (0.4% and 2.0%, respectively) but much higher during the following years. Such taxa build up a seed bank waiting for the right conditions to sprout, with diaspore morphology and diaspore weight determining how long the seeds are able to survive. A higher weight and a more compact seed coat increase the chance for successful germination after some years of dormancy. Those differences in germination behavior are reflected by the species frequencies in the glacier forelands studied. Species which are able to sprout immediately after reaching the glacier foreland show a swift increase in individual numbers and ground cover (e.g., *O. digyna, G. reptans, C. uniflorum, A. alpina*), while others with low germination success without interim dormancy such as *L. alpina* are significantly underrepresented [17].

Once diaspores have germinated, the next obstacle is to survive the juvenile stage, which is a particularly sensitive phase and characterized by high mortality rates due to different potential threats [3, 5, 26, 35], which should be discussed next. Despite low contents of organic material, nitrogen, and phosphor, nutrient matter does not seem to be responsible for seedling mortality in glacier forelands. Just after ice retreat, the substrate has sufficient nutrients by atmospheric dust and N-depositions to instantly allow plant growth [28, 36–39]. Algae, cryptogams, and mosses might be involved but obviously do not play an important role in site melioration [17, 21, 40]. More important for the establishment of diaspores than nutrient matters are probably the prevailing substrate conditions. Glacier forelands are commonly regarded as unconsolidated, instable ground with a high amount of coarse rocks impeding colonization by plants. The permanent plots on Goldbergkees and Lenksteinferner locally feature a high amount of coarse material, but situated on more or less leveled ground this is remarkably solid [17]. Soil frost activity (solifluction, cryoturbation) is effectively suppressed by a long-lasting snow cover for 8–9 months and only rare freezing events during the snow-free season (late June/early July to late September/early October). Thus substrate instability cannot be assumed a universal factor for seedling mortality in alpine glacier forelands. Also, a high amount of coarse boulders is no obstacle for colonization, a minimum of fine grained substrate provided for rooting and water supply. In fact, larger rocks provide safe sites with microclimatically more favorable conditions (shorter snow cover duration, pronounced warming, etc.), and a lack of such safe sites significantly enhances seedling mortality and slows down early colonization within glacier forelands [11]. Despite a high amount of precipitation, reduced evapotranspiration, and additional water supply by melting snow and ice, desiccation of the coarse-grained substrates could be another important reason for mortality in glacier forelands, in particular during seedling and early development stage [4, 41, 42]. In particular under longer-lasting drought phases and/or reduced snow cover related to climate warming, desiccation might become a more important issue in the future. Concerning soil temperatures multiyear measurements in the root horizon of plants (−10 cm) show rather mild conditions within glacier forelands [17]. Despite high inter- and intra-annual variations (~50%) which are expressed primarily in the length of vegetation period and temperature sums, mean temperatures between 6 and 10°C were recorded during the snow-free period in the glacier forelands of Lenksteinferner and Goldbergkees. These are higher temperatures than at tree line, where the trees make themselves a cold root horizon by shadowing effects [43]. Freezing temperatures within the root horizon in the glacier forelands occur but are rare during the snow-free season. While seeds are rather unsusceptible to moderate freezing, seedlings are not [6]. Species investing primarily in aboveground biomass are particularly at risk, while those that invest mainly in below-ground biomass during the first year (e.g., *O. digyna*) are less vulnerable, show lower mortality rates, and are represented by higher individual numbers and ground cover values [17, 25]. Besides desiccation and freezing during the vegetation period, the winter months are the second crucial phase for the survival of seedlings. A snow cover lasting for too long can prevent successful establishment and carbon gain; if snow cover removal is too early, the risk of freezing damage to the seedlings is high, and in addition, periglacial processes may mechanically negatively affect the roots. In the glacier forelands of Goldbergkees and Lenksteinferner, seedling mortality in general is low, indicating that none of the mentioned potential threats is common or at least was common during the study period between 2005 and 2015.

3.3. Step 3 in primary succession: grow up and spread

The third important step for a successful plant colonization of new ground is grow up and spread [18]. Many of the early colonizers are long-lived taxa, with a life expectancy of up to 50 years, in clonal and cushion plants even more [42]. Once established, plants occupy their sites for decades [44, 45], unless the site conditions change fundamentally. With water and nutrient supply ensured, established plants grow and gain ground cover under almost uncompetitive conditions. Like all perennial plants, also the high-elevation specialist alternates between phases of growth and phases of reproduction, which are subject to seasonal cycles [46]. Day length and/or a priori low-temperature period during winter ("vernalization") control the right timing of flowering. While after the cold stimulus and during early summer investment is mainly in reproductive plant parts, after fruiting biomass increase is again paramount (see Ref. [47] for *A. alpina*). Simultaneously with growth new individuals establish from both external seed sources and diaspore-bearing individuals on the sample sites itself. A synchronous operation of steps 1, 2, and 3 side by side is pushing forward succession. In consequence, a positive logarithmic or even exponential increase of species numbers, individual numbers, and ground coverage emerges as soon as the established individuals produce diaspores, which commonly happens the second year after establishment [25]. As most seeds are deposited in the immediate surrounding of the mother plant (leptokurtic diaspore dispersal behavior!), this direct diaspore input is superior to long-distance dispersal [5, 25, 27]. In addition, the ability for self-pollination in many taxa enhances the reproductive success, albeit at the expense of genetic variability.

Besides seed rain also vegetative propagation of capable species is relevant for increasing ground cover values and individual numbers of established plants [42, 45]. One of those species that perform both generative and vegetative propagations is the stolon-producing *G. reptans* (see **Figure 4**). The downside of reduced genetic variability is compensated by spread even in unfavorable years prohibiting generative propagation [48]. Another way of vegetative (clonal) reproduction is performed by *P. alpina* (**Figure 4**) which is producing bulbils in its pseudo-viviparous form. The development of genetically identical daughter plants instead of seeds is triggered by unfavorable site conditions and thus becomes more important under adverse conditions with higher elevation [42]. In the glacier forelands, the pseudo-viviparous form is much more common than the normal seed-producing form. The daughter plants are photosynthetically active already on the mother plant and after release are dispersed by wind. In doing so, a faster and more successful establishment within the glacier forelands is guaranteed compared to the development of diaspores with all the uncertainties during establishment. In preserving the genetic information, this strategy could even be a selective advantage for *P. alpina* in comparison to the non-viviparous form [49].

The permanent plot studies show a very dynamic colonization of recently deglaciated ground. Vegetation dynamics in general are governed by the relative favor or disfavor of a site, which is a function of many different interrelated factors such as elevation, exposure, snow cover duration, continentality, existence of safe sites, seed sources, etc. Despite differences in the absolute values, in general, a swift fill-up of empty niches is taking place during the first decade of the permanent plot study, promoted by the growth of established individuals as well as by a continuous colonization by new individuals. These are invasive populations according to Ref. [50] with young individuals prevailing. A more uniform ("Gaussian") distribution with many species of intermediate age and few young and old ones, which would indicate stable populations (see Ref. [45]), is not yet reached. Several species invest most of the resources in generative reproduction and are able to create persistent diaspore banks. Many of the early colonizers also show ruderal characteristics [51] such as fast increment, the ability for self-pollination, or anemochorous dispersal. Just like their lowland counterparts, alpine ruderals are able to colonize sites with a high disturbance frequency, unlike, however, is their much greater life expectancy [42, 45]. Most of the species encountered in the sample sites are far from their maximum age, explaining the low number of losses in the repeated surveys. If dropouts occur, they are compensated by newly established individuals of the same species, expressed in a general increase of groundcover and individual numbers. One notable exception is *A. alpina* with a life expectancy of only a couple of years. In both glacier forelands, this species is regularly present with diebacks during resurveys. *A. alpina*, however, as self-pollinating species [47] produces a high number of diaspores able for immediate germination and thus a high number of seedlings every year. The short life cycles of *A. alpina* causing a continuous dieback of individuals keep increase of individual numbers moderate compared to more long-lived species such as *O. digyna*, *C. uniflorum*, or *S. bryoides*. Only locally temporary setbacks in the overall vegetation development by disturbances are apparent [17]. Most common is the displacement of meltwater runoff over-pouring the sample sites for a while. The overdose of (cold) water, probably combined with a higher frequency and intensity periglacial processes in the substrate (solifluction, cryoturbation, needle ice, etc.) negatively impacts the life processes of the plants, which is expressed in diebacks of quite a number of individuals of different species in such cases. When the melt-out stops or the runoff is displaced again, progressive developments reemerge.

4. Long-term vegetation development in glacier forelands as indicated by chronosequences

To evaluate vegetation dynamics in glacier forelands on a temporally larger scale, chronosequences are commonly employed. Despite some shortcomings (see above) chronosequences are helpful to hypothesize about long-term vegetation development and offer a good baseline to be corroborated or dismissed by long-running permanent plot studies. **Figure 5** exemplifies gradual vegetation changes with time for the chronosequence in the glacier foreland of Goldbergkees. Based on the floristic composition and structural attributes, different successional stages can be identified for both chronosequences surveyed: a pioneer stage, an early

Figure 5. Aspects of sample sites along the chronosequence in the glacier foreland of Goldbergkees (the central site out of three per level) illustrate the gradual vegetation change with time. Time since melt-out is 2 years (a), 4 years (b), 15 years (c), 25–30 years (d), 55 years (e), 85 years (f), 120 years (g), and 155 years (h).

successional stage and a later successional stage, which might be further subdivided in a grass-dwarf shrub-phase and a shrub phase. A superordinate species pool characteristic for the siliceous Eastern Alps causes a general similarity between glacier forelands of the Central Eastern Alps (and also for the two presented here); however, the "floristic character of the surroundings" [19] depending on local site conditions (elevation, exposure, topography, etc.) is responsible for some discreetness [17].

The PCAs in **Figure 6** portray the floristic similarity between samples (the closer located within the ordination space, the higher the similarity) as well as changes in groundcover and life form composition during succession for the two chronosequence studies. The pioneer stage includes sites deglaciated for up to 20 years (A- to C-sites on Goldbergkees, B-sites on Lenksteinferner, A-sites on Lenksteinferner are missing as those three selected were still devoid of vegetation in 2007; see **Figure 6**), and vegetation development basically reflects the situation on the permanent plots one decade after deglaciation presented above. Substrate is blocky without any signs of initial soil development. Mineral nutrient supply is guaranteed via sediment input by wind and melting glaciers as well as by dry and wet N-deposition [36–39], creating a first "natural manuring." As already revealed by the permanent plot studies, the chamaephytes *C. uniflorum*, *C. cerastoides*, *S. bryoides*, and *A. alpina*; herbs such as *O. digyna*, *Sagina saginoides*, *V. alpina*, and *Cardamine resedifolia*; the grass *P. laxa*; as well as mosses belong to the early colonizers in both glacier forelands. All of the vascular plants are anemochorous species carried to the glacier foreland by valley winds from the surroundings, where they are able to establish without interspecific competition. Within the first two decades, 20–30 different taxa appear on the new ground. Which species accompany the mentioned early colonizers depends on the seed sources in the surrounding and the local site conditions. In general, the pioneer stage is characterized by a high degree of randomness. Ground cover is low with <2% on the sites sampled. The early successional stage encompasses sites between 20 and 60 years of age (D- and E-sites on Goldbergkees, C- and D-sites on Lenksteinferner; see **Figure 6**). Some species already sparsely present within the pioneer stage gain importance (e.g., *L. alpina*, *G. reptans*, *Gnaphalium supinum*, as well as the grasses *A. rupestris* and *P. alpina*, again predominantly in the viviparous form), and many additional species join the sites. In total, 30 different species of vascular plants were recorded for the early successional stage on Goldbergkees and 31 on Lenksteinferner. Ground cover increases to values around 10% on sites deglaciated for roughly half a century, and dwarf shrubs are the predominant life form (**Figure 6**). Most of the early colonizers are still present with high frequency and/or abundance; on sites older than half a century; however, a continuous influx of additional species segregates the later successional stage (F- to H-sites on Goldbergkees, F- to J-sites on Lenksteinferner) and induces a generally higher dissimilarity between sites than earlier stages (**Figure 6**). One very common species is *Euphrasia minima*, one of the few therophytes involved in primary succession in alpine glacier forelands. Species numbers increase to well over 40, and mean ground cover is around 60%. Increasing ground cover and species richness intensify competition and eliminate some pioneers weak in competition. The later successional stage can be further divided into a grass-dwarf shrub phase (present on Goldbergkees) and a shrub phase (present on Lenksteinferner). In the former, the carpet-like dwarf willows *Salix herbacea* and *Salix retusa* are very common. The shrub phase is characterized by a higher groundcover of more upright-growing shrubs such as *Rhododendron ferrugineum* and different willow species (on Lenksteinferner *Salix appendiculata* and *Salix breviserrata*). In addition, about 120–150 years after deglaciation, the first conifer taxa such as *Juniperus communis* ssp. *nana*, *L. decidua*, or *Picea abies* are present in the sample sites on Lenksteinferner (**Figure 6**).

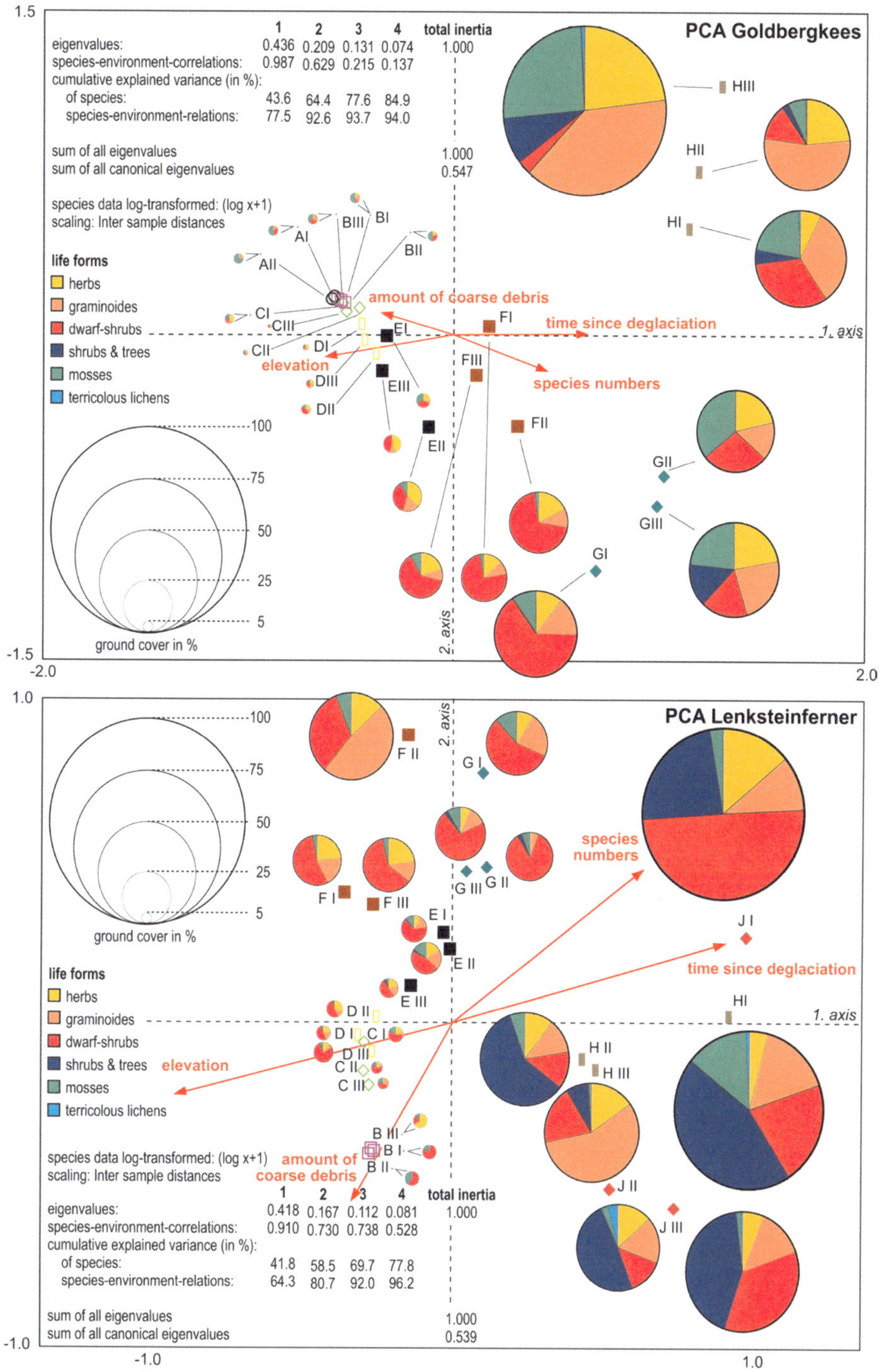

Figure 6. Principal component analyses (PCAs) based on species data for the chronosequence studies in the glacier forelands of Goldbergkees (above) and Lenksteinferner (below). The pies indicate life form composition and total ground cover for all samples. Where necessary (A, B, and C on Goldbergkees, B on Lenksteinferner) pies are zoomed for better reading.

Succession on new ground is commonly reflected by an increase of species numbers and ground cover, at least until a certain point [52] (see **Figure 6**). Species diversity of a particular successional stage is not so much triggered by elevation, rather by the vegetation belt in which it is located. For instance, on Goldbergkees sites being deglaciated for one and a half century are located within the alpine belt and exhibit less species than the same-aged sites on Lenksteinferner which are—despite higher absolute elevation—located close to the treeline ecotone allowing for an association of subnival, alpine, and subalpine elements. While the increase of species numbers shows a more negative logarithmic behavior, the development of ground cover is positive logarithmic, i.e., despite a swift increase of species numbers during the pioneer and early successional stages, ground cover values lag behind during the first decades (see **Figure 7**)—a pattern already observable during the first decade of the permanent plot study. Approximately half a century after deglaciation, a speedup in ground cover becomes apparent. This increase is not always continuous; rather disturbances such as mudflows, relocation of glacial runoff, avalanches, etc. can throw back succession to an earlier stage, as displayed by the decrease in both species numbers and ground cover values on the 90-year-old G-sites on Lenksteinferner (see **Figure 7**).

Primary succession in glacier forelands is a process that always occurred when glaciers receded, whether in postglacial times, after the LIA, or today with recent climate warming. While the general processes of primary succession were basically always the same, the circumstances controlling these processes may differ between today and the past. Recent studies, at least, found primary succession within glacier forelands of the Alps to be accelerated,

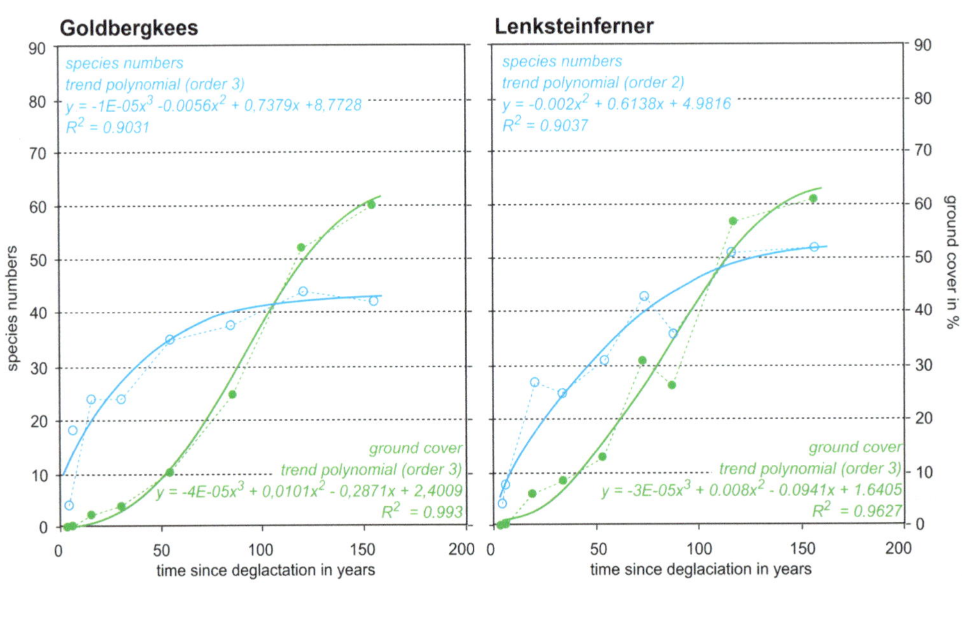

Figure 7. Development (mean out of three samples) of species numbers (blue open circles) and ground cover (green closed circles) along the chronosequences in the glacier forelands of Goldbergkees (left) and Lenksteinferner (right).

http://dx.doi.org/10.5772/intechopen.69479

most likely due to climate warming (e.g., Refs. [24, 53]). As a complete list of species present in the foreland of Lenksteinferner was already published for the early twentieth century [19], this glacier foreland offers the great opportunity to compare these historic data with those collected roughly one century later (i.e., those presented here). Ref. [54] employed these two spatiotemporally different data sets to address the question whether primary succession of plants in glacier forelands today differs from the past concerning the dynamics of colonization, the plant species involved, and their respective biological traits. The main outcome of this study was that even if additional species occur and the colonization apparently is faster today compared to the past, fundamental differences concerning the floristic inventory, the biological traits, or the colonization strategies of the early colonizers due to climate change do not exist. This is apparently a consequence of a compensation of climate warming during the twentieth century by the shift of the glacier terminus to a higher elevation. The vertical shift of the glacier snout of Lenksteinferner between the early twentieth and early twenty-first century amounts approximately 300 m in elevation. Assuming a mean adiabatic temperature lapse rate of −0.57 K/100 m, mean annual temperatures between the two elevational levels differ by 1.7 K [54]. This value corresponds quiet well to the magnitude of climate warming between the two sampling dates, and therefore, almost identical thermal conditions can be assumed for the recent glacier foreland (at higher elevation but affected by climate warming) and the one at the beginning of the last century (at lower elevations but under colder climate). As a shift in elevation of glacier termini during recession is a common issue, such compensation effects can be assumed to be a widespread determinant for succession in glacier forelands of the Alps (and elsewhere). Glaciers which terminate in flat glacial valleys (e.g., Morteratsch glacier in the Swiss Engadine [53]) may react differently, as compensation effects of elevation change by climate warming are lacking and the increasing temperatures may immediately affect plant colonization and vegetation dynamics in glacier forelands, allowing thermophilous species of lower elevations to participate in primary succession.

5. Conclusion

Studies on plant succession are highly important tasks, not only for the general understanding of the colonization on newly created surfaces but also for providing insight for rehabilitation measures on disturbed ecosystems in general. The combined use of permanent plots and chronosequences interrelates the benefits of the two different approaches while reducing the respective drawbacks. Thus, cautious interpretation allows for the deduction of trends in vegetation dynamics on larger time scales. The permanent plot studies reveal a highly dynamic vegetation development in recently deglaciated glacier forelands with the first plant individuals appearing soon after deglaciation. A surprising fact is how swift the increase of species and individual numbers within the first decade after deglaciation is taking place, and it seems that mutualistic effects are important, while competition does not play a major role at present. Inter- and intraspecific competition becomes more effective later on during succession and in particular becomes apparent one century or more after deglaciation with the dropout of several early colonizers. The high persistency documents that not only pioneer species are involved in colonization of bare ground but early and late-successional taxa as well. Even if conclusions of chronosequence studies have to be drawn carefully as other factors than site age might be similarly responsible for the vegetation development encountered, they allow hypothesizing about the future development of recently deglaciated glacier forelands. Whether the trends deduced by chronosequences on the permanent plots in fact appear remains to be seen when the long-term monitoring will be continued over the next decades.

Author details

Thomas Fickert

Address all correspondence to: thomas.fickert@uni-passau.de

Physical Geography, University of Passau, Passau, Germany

References

[1] Ellenberg H. Vegetation Ecology of Central Europe. Cambridge: Cambridge University Press; 1988. p. 731

[2] Pickett STA. Space-for-time substitution as an alternative to long-term studies. In: Likens GE, editor. Long-term Studies in Ecology: Approaches and Alternatives. New York: Springer; 1989. pp. 110-135

[3] Matthews JA. The Ecology of Recently-deglaciated Terrain. A Geoecological Approach to Glacier Forelands and Primary Succession. Cambridge: Cambridge University Press; 1992. p. 386

[4] Innes JL. Lichenometry. Progress in Physical Geography. 1985;**9**:187-254

[5] Jones CC, del Moral R. Dispersal and establishment both limit colonization during primary succession on a glacier foreland. Plant Ecology. 2009;**204/2**:217-230

[6] Marcante S, Sierra-Almeida A, Spindelböck JP, Erschbamer B, Neuner G. Frost as a limiting factor for recruitment and establishment of early development stages in an alpine glacier foreland? Journal of Vegetation Science. 2012;**23/5**:858-868

[7] Connell JH, Slatyer RO. Mechanisms of succession in natural communities and their role in community stability and organization. The American Naturalist. 1977;**111/982**:1119-1144

[8] Olofsson J, Moen J-L, Oksanen L. On the balance between positive and negative plant interactions in harsh environments. Oikos. 1999;**86**:539-543

[9] Callaway RM, Brooker RW, Choler P, Kikvidze Z, Lortiek CJ, Michalet R, Paolini L, Pugnaireq FI, Newingham B, Aschehoug ET, Armasq C, Kikodze D, Cook BJ. Positive interactions among alpine plants increase with stress. Nature. 2002;**417**:844-848

[10] Dullinger S, Kleinbauer I, Pauli H, Gottfried M, Brooker R, Nagy L, Theurillat J-P, Holten JI, Abdaladze O, Benito J-L, Borel J-L, Coldea G, Ghosn D, Kanka R, Merzouki A, Klettner C, Moiseev P, Molau U, Reiter K, Rossi G, Stanisci A, Tomaselli M, Unterlugauer P, Vittoz P, Grabherr G. Weak and variable relationships between environmental severity and small-scale co-occurrence in alpine plant communities. Journal of Ecology. 2007;**95**:1284-1295

[11] Erschbamer B, Niederfriniger Schlag R, Winkler E. Colonization processes on a central alpine glacier foreland. Journal of Vegetation Science. 2008;**19/6**:855-862

[12] Raunkiaer C. The Life-forms of Plants and Statistical Plant Geography. Oxford: Oxford University Press; 1934. p. 632

[13] Müller-Schneider P. Verbreitungsbiologie der Blütenpflanzen Graubündens. Zürich: Veröffentlichungen des Geobotanischen Instituts der Eidg. Techn. Hochschule, Stiftung Rübel, 85; 1986. p. 263

[14] Fischer MA, Adler W, Oswald K. Exkursionsflora für Österreich, Liechtenstein und Südtirol. Linz: Land Oberösterreich, Biologiezentrum der OÖ Landesmuseen; 2005. p. 1392

[15] Böhm R, Schöner W, Auer I, Hynek B, Kroisleitner C, Weyss G. Gletscher im Klimawandel – Vom Eis der Polargebiete zum Goldbergkees in den Hohen Tauern. Wien: ZAMG; 2007. p. 111

[16] Damm B. Die Entwicklung der Vergletscherung in der Rieserfernergruppe (Tirol) seit dem Hochstand von 1850. Der Schlern. 1997;**71/10**:593-618

[17] Fickert Th. Zur Bedeutung von Dauerbeobachtungsstudien in der Sukzessionsforschung-Zwei Fallstudien zur Primärsukzession in Gletschervorfeldern in den Ostalpen und zur Sekundärsukzession sturmgestörter Mangrovenwälder in Honduras. Passauer Schriften zur Geographie H. 29, Passau: Selbstverlag Fach Geographie der Universität Passau; 2016. p. 158

[18] Bradshaw A. Understanding the fundamentals of succession. In: Miles J, Walton DWH, editors. Primary Succession on Land. Oxford, Boston: Blackwell Scientific Publications; 1993. pp. 1-3

[19] von Klebelsberg R. Das Vordringen der Hochgebirgsvegetation in den Tiroler Alpen–Eine alpin-pflanzengeographische Studie. Österreichische Botanische Zeitschrift. 1913;**63/5** & **63/6**:77-186 & 241-254

[20] Lüdi W. Beobachtungen über die Besiedlung von Gletschervorfeldern in den Schweizeralpen. Flora. 1958;**146**:387-407

[21] Richard JL. Dynamique de la végétation au bord du grand glacier d'Aletsch (Alpes Suisses). In: Schmidt W, editor. Sukzessionsforschung. Berichte der Internationalen Symposien der Internationalen Vereinigung für Vegetationskunde (Rinteln 1973), Cramer, Vaduz; 1975. pp. 189-209

[22] Burga CA, Krüsi B, Egli M, Wernl M, Elsener S, Ziefl M, Fischer T, Mavrisa M. Plant succession and soil development on the foreland of the Morteratsch glacier (Pontresina, Switzerland): Straight forward or chaotic? Flora. 2010;**205**:561-576

[23] Zollitsch B. Die Vegetationsentwicklung im Pasterzenvorfeld. Wissenschaftliche Alpenvereinshefte. 1969;**21**:267-290

[24] Cannone N, Diolaiuti G, Guglielmin M, Smiraglia C. Accelerating climate change impacts on alpine glacier forefield ecosystems in the European Alps. Ecological Applications. 2008;**18/3**:637-648

[25] Stöcklin J, Bäumler E. Seed rain, seedling establishment and clonal growth strategies on a glacier foreland. Journal of Vegetation Science. 1996;**7**:45-56

[26] Erschbamer B, Kneringer E, Niederfriniger Schlag R. Seed rain, seed bank, seedling recruitment, and survival of seedlings on a glacier foreland in the Central Alps. Flora. 2001;**196/4**:304-312

[27] Tackenberg O, Stöcklin J. Wind dispersal of alpine plant species: A comparison with lowland species. Journal of Vegetation Science. 2008;**9/1**:109-118

[28] Caccianiga M, Luzzaro A, Pierce S, Ceriani RM, Cerabolini B. The functional basis of a primary succession resolved by CSR classification. Oikos. 2006;**112**:10-20

[29] Becker T, Dierschke H. Primärsukzession im Gletschervorfeld des Obersulzbachkees (Hohe Tauern, Österreich), eine Zeitreihe über fast 150 Jahre. Tuexenia. 2005;**25**:111-139

[30] Fickert T, Friend D, Grüninger F, Molnia B, Richter M. Did debris-covered glaciers serve as Pleistocene refugia for plants? A new hypothesis derived from observations of recent plant growth on glacier surfaces. Arctic, Antarctic, and Alpine Research. 2007;**39/2**:245-257

[31] Caccianiga M, Andreis C, Diolaiuti G, D'Agata C, Mihalcea C, Smiraglia C. Alpine debris-covered glacier as habitat for plant life. The Holocene. 2011;**21/6**:1011-1020

[32] Schwienbacher E, Erschbamer B. Longevity of seeds in a glacier foreland of the Central Alps – A burial experiment. Bulletin of the Geobotanical Institute ETH. 2001;**68**:63-71

[33] Schwienbacher E, Marcante S, Erschbamer B. Alpine species seed longevity in the soil in relation to seed size and shape—A 5-year burial experiment in the Central Alps. Flora. 2010;**205**:19-25

[34] Schwienbacher E, Navarro-Cano JA, Neuner G, Erschbamer B. Seed dormancy in alpine species. Flora. 2011;**206**:845-856

[35] Marcante S, Winkler E, Erschbamer B. Population dynamics along a primary succession gradient: Do alpine species fit into demographic succession theory? Annals of Botany. 2009;**103**:1129-1143

[36] Haselwandter K, Hofmann A, Holzmann HP, Read DJ. Availability of nitrogen and phosphorus in the nival zone of the Alps. Oecologia. 1983;**57**:266-269

[37] Chapin III FS, Walker LR, Fastie CL, Sharman LC. Mechanisms of primary succession following deglaciation at Glacier Bay, Alaska. Ecological Monographs. 1994;**64/2**:149-175

[38] Heer C, Körner Ch. High elevation pioneer plants are sensitive to mineral nutrient addition. Basic and Applied Ecology. 2002;**3**:39-47

[39] Hiltbrunner E, Schwikowski M, Körner Ch. Inorganic nitrogen storage in alpine snow pack in the Central Alps (Switzerland). Atmospheric Environment. 2005;**39**:2249-2259

[40] Nagy L, Grabherr G. The Biology of Alpine Habitats. Oxford: Oxford University Press; 2009. p. 376

[41] Urbanksa CM, Schütz M. Reproduction by seed in alpine plants and revegetation research above timberline. Botanica Helvetica. 1986;**96**:43-61

[42] Körner Ch. Alpine Plant Life. Berlin, Heidelberg, New York: Springer; 2003. p. 344

[43] Körner Ch. Alpine Treelines. Basel, Heidelberg, New York, Dodrecht, London: Springer; 2012. p. 220

[44] Grabherr G, Gottfried M, Pauli H. Long-term monitoring of mountain peaks in the Alps. In: Burga CA, Kratochwil A, editors. Biomonitoring: General and Applied Aspects on Regional and Global Scales, Tasks for Vegetation Science 35. Dordrecht, NL.: Kluwer; 2001. pp. 153-177

[45] Erschbamer B, Retter V. How long can glacier foreland species live? Flora. 2004;**199/6**: 500-504

[46] Battey NH. Aspects of seasonality. Journal of Experimental Botany. 2000;**51/352**:1769-1780

[47] Wang R, Farrona S, Vincent C, Joecker A, Schoof H, Turck F, Alonso-Blanco C, Coupland G, Albani MC. PEP1 regulates perennial flowering in *Arabis alpina*. Nature. 2009;**459**:423-427

[48] Weppler T, Stoll P, Stöcklin J. The relative importance of sexual and clonal reproduction for population growth in the long-lived alpine plant *Geum reptans*. Journal of Ecology. 2006;**94**:869-879

[49] Pierce S, Stirling CM, Baxter R. Pseudoviviparous reproduction of *Poa alpina* var. *vivipara* L. (Poaceae) during long-term exposure to elevated atmospheric CO_2. Annals of Botany. 2003;**91/6**:613-622

[50] Rabotnov TA. Differences between fluctuations and successions. In: Knapp R, editor. Vegetation Dynamics. Handbook of Vegetation Science 8. The Hague: Dr. W. Junk b.v., Publishers; 1974. pp. 19-24

[51] Grime JP. Vegetation classification by reference to strategies. Nature. 1974;**250**:26-31

[52] Andreis C, Caccianiga M, Cerabolini B. Vegetation and environmental factors during primary succession on glacier forelands. Plant Biosystems. 2001;**135/3**:295-310

[53] Burga CA. More and quicker new land for pioneer plants and faster plant settlement—14 years plant monitoring of the glacier foreland Morteratsch (2000 m a.s.l., Pontresina, Upper Engadine, Switzerland). In: Frei ER, Stöckli V, Rixen C, Wipf S, editors. Faster, Higher, More? Past, Present and Future Dynamics of Alpine and Arctic Flora under Climate Change. Abstracts. Intern. Conference; September 22-25, 2013; Bergün, Switzerland. Birmensdorf: Swiss Federal Institute for Forest, Snow and Landscape Research WSL; 2013

[54] Fickert T, Grüninger F, Damm B. Klebelsberg revisited: Did primary succession of plants in glacier forelands a century ago differ from today? Alpine Botany. 2017;**127**:17-29. DOI: 10.1007/s00035-016-0179-1

Chapter 7

Animal Successional Pathways for about 200 Years Near a Melting Glacier: A Norwegian Case Study

Sigmund Hågvar, Mikael Ohlson and Daniel Flø

Additional information is available at the end of the chapter

http://dx.doi.org/10.5772/intechopen.68192

Abstract

Here, we explore 200 years of arthropod succession by using dated moraines in a Norwegian glacier foreland. Surface active beetles (Coleoptera) and spiders (Aranea) were sampled by pitfall trapping, and springtails (Collembola) and mites (Acari) were extracted from soil samples. Newly deglaciated ground was rapidly colonised by a mixture of generalists and specialists, with various life strategies. Interestingly, the pioneer community was fed by three 'invisible' food sources: biofilm with terrestrial diatom algae, tiny pioneer mosses and chironomid midges whose larvae were pond-living and used ancient carbon that was released by the melting glacier as an energy source. The true 'super-pioneers' were biofilm-eating springtails, which tracked the melting ice edge closely. Most species of beetles and springtails colonised within 80 years, while spiders and oribatid mites needed a longer time span to colonise. Topography influenced the succession pattern. Among both surface-living macroarthropods and soil-living microarthropods, we distinguished between a 'dry' and a 'wet' successional pathway with different community structure. Most arthropod species persisted after colonisation, but certain species preferring open space or low temperature were gradually excluded. Comparisons are made with botanical succession. Sampling methods, material size, and taxonomic resolution were considered critical factors when studying arthropod succession.

Keywords: succession, beetles, spiders, springtails, mites, glacier foreland, moisture, alternative successional pathways, geo-ecology

1. Introduction

Due to climate change, glaciers are shrinking worldwide [1–3]. Simultaneously, 'waves' of different organisms try to colonise the newly exposed land. Glacier forelands give unique possibilities for studying primary succession. Instead of monitoring changes within a fixed plot over time, which indeed would be a very time-demanding approach, successions can be described by studying sites with known ages. Such a gradient in the terrain, where space is used as a substitute for time, is called a chronosequence [4, 5].

Most studies in glacier forelands have dealt with plant succession, and a thorough and long-lasting one has been performed near the glacier Storbreen in Norway [4]. A main conclusion is that age alone cannot predict the plant community. Local variations in microtopography, moisture, nutrients, substrate and exposure contribute in shaping the species composition. Instead of ending up with a 'monoclimax', the succession produces a 'polyclimax' with a mosaic of plant communities. A 'bulk' succession related only to the age of the ground contains 'noise' from a mixture of successional pathways. It has been argued for a 'geo-ecological' view on primary succession, where both biotic and abiotic factors were taken into considerations [4]. A recent study from Nigardsbreen foreland in Norway confirmed the modifying effect of microtopography on the floral succession [6].

Studies on animal succession near receding glaciers are fewer, and are mainly focusing on arthropods. In addition to the present case study, there are studies on arthropod succession in glacier forelands from the Alps [5, 7–11], from Svalbard [12–15], from Iceland [16] and from Norway [17–23].

In the following presentation, we have adopted the geo-ecological perspective. In other recent studies of invertebrate successions in Norwegian glacier forelands, a geo-ecological perspective has been successfully applied, when comparing succession patterns at different altitudes and climatic conditions [19, 20, 22, 23].

Hardangerjøkulen glacier in Southern Norway has been receding since the end of 'the little ice age' for about 250 years ago. The melting rate has been especially high during the last two decades, with about 20-m retreat yearly at one glacier snout near Finse (Midtdalsbreen). We have good data on earlier positions of the ice edge in this glacier foreland due to dated moraines. Since 2001, extensive zoological studies have been performed here to describe and understand arthropod succession patterns (**Figure 1**). These studies include soil-living micro-arthropods [24, 25], surface active beetles, spiders and harvestmen [26, 27], aerial transport of arthropods [28], studies on ancient carbon released by the glacier [29, 30], food choice of pioneers [31], as well as a special focus on early succession [32].

Time has come to combine these fragments into a holistic story about animal succession near a melting glacier. In addition to summing up the main results from these nine papers, the present syntheses will discuss some general aspects of succession:

a. *Comparison between botanical and zoological succession*

- Are there alternative succession patterns for arthropods, in the same way as there are alternative succession patterns for vegetation [4–6, 33]?
- Is there a strong progressive succession of arthropods on terrain ages of 20–50 years, as in plants [33]?
- Does animal and botanical successions differ in their early phases?

Figure 1. Sampling of soil for the extraction of microarthropods (springtails and mites). This snow bed habitat was 180 years old and situated 1 km from the glacier (plot no. 18). It had a well-developed *Salix herbacea* vegetation and a 2-cm thick organic soil layer. Microtopography in the surroundings created gradients from dry ridges to moist depressions.

b. *Questions about zoological succession*

- Do most arthropod species tend to persist after colonising [19, 20, 22, 23]?
- Is a geo-ecological perspective fruitful and relevant when considering mechanisms of facilitation and inhibition in zoological successions?
- Does arthropod succession pattern differ between surface-active macroarthropods and soil-living microarthropods?
- Is soil fauna succession in *Salix herbacea* snow bed vegetation related to the gradual development of an organic layer?

c. *Question about methods*

- Are sampling methods, material size and taxonomic resolution critical factors when studying arthropod succession?

2. Materials and methods

2.1. Study site

The study site was situated close to the 73-km^2 large Hardangerjøkulen glacier in central Southern Norway, between 1200 and 1400 m a.s.l., in the treeless low- and mid-alpine zone. **Figure 2** shows the foreland of a northern glacier snout named Midtdalsbreen (60°34′N, 7°28′E).

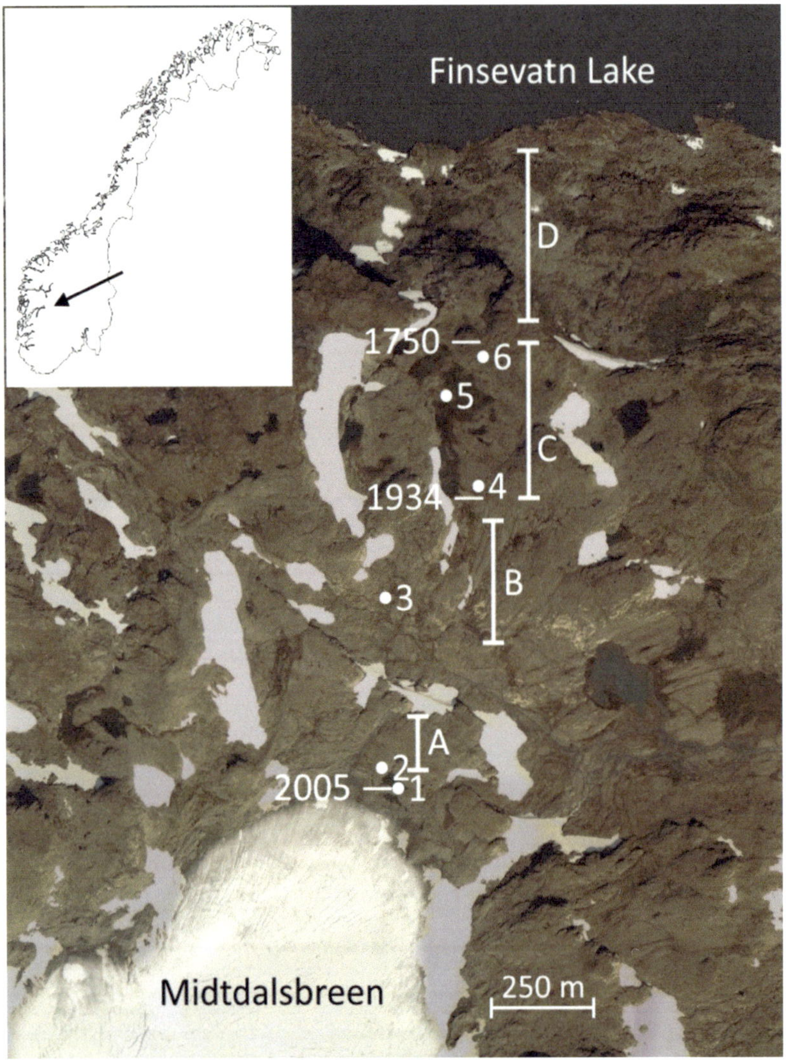

Figure 2. Aerial photograph of the glacier foreland, showing the position of moraines from 1750, 1934 and 2005. Beetles, spiders and harvestmen were sampled on the six numbered plots with the following ages: 3, 40, 63, 79, 170 and 205 years. Springtails and mites were sampled from eight plots in zone A (32–48 years), five plots in zone B (52–66 years), seven plots in zone C (72–227 years) and five plots in zone D (10,000 years). The small map shows the position of the foreland in Southern Norway. Modified from Ref. [26].

2.2. Microarthropod sampling

For the study of soil-living microarthopods, which are springtails (Collembola) and mites (Acari), we chose to keep the vegetation factor as constant as possible. All soil samples were taken in *S. herbacea* vegetation, which was found throughout the gradient. This tiny shrub belongs to the pioneer plants and shows no preference for snow cover on relatively young terrain [34, 35]. However, after about 70 years, it is mainly restricted to patches where snowmelt is late (the so-called snow beds), where it forms rather continuous carpets. Plots 1–8 (zone A) were 32–48 years old, plots 9–13 (zone B) were 52–66 years old and plots 14–20 (zone C) were 72–227 years old. Plot nos. 21–25 (zone D) were outside the 1750 moraine which mark the end of the 'little ice age' in Norway, so these five plots had an age of about 10,000 years [24]. In each of the 25 study plots, microarthropods were extracted from 10 to 16 soil cores, 3 cm deep and with a surface area of 10 cm^2 [24].

2.3. Macroarthropod sampling

A different sampling strategy was chosen for surface-active macroarthropods, which were beetles (Coleoptera), spiders (Aranea) and harvestmen (Opiliones). Here, we aimed at collecting as many species as possible at each age, by covering a span of vegetation types. Pitfall traps were used at six sites with the following ages: 3, 40, 63, 79, 170 and 205 years. Twenty traps with a diameter of 6.5 cm were operated at each site and emptied every 2 weeks during two snow-free seasons (2007 and 2008) [26]. Traps were usually distributed in a topographic gradient from dry, lichen-dominated vegetation via an *Empetrum hermaphroditum* heath, to moist snow bed. In a nearby foreland of the same glacier (Blåisen glacier snout), these plant communities were characteristic products of succession [35]. Vegetation and the degree of cover were noted around each trap, a number of soil moisture data were taken, and catches from each trap and period were kept separate. Pitfall traps measure surface activity and not density, but they catch a high number of species and may be used in comparison between sites.

Figure 3. A small pond, 8 years old, in which larvae of chironomid midges assimilated ancient carbon from the sediments. From Ref. [30].

Aquatic invertebrates, for instance larvae of Chironomidae midges, were sampled from young ponds using a sieve. **Figure 3** shows a pond on an 8-year-old moraine.

2.4. Sticky traps and fallout traps

We performed extensive sampling of airborne arthropods on 3–6-year-old ground on the 2005 moraine [28]. Two types of traps were used: sticky traps and fallout traps. The sticky traps were placed on poles, up to 1-m height, and turned towards different directions. Fallout traps had their rim 5 or 11 cm above the ground to prevent surface-active arthropods to drop into them (**Figure 4**).

2.5. Gut content analyses

The food choice of different species was studied by analysing their gut contents under the microscope. Crop and gut contents of beetles and harvestmen were dissected out and spread on slides, embedded in glycerol. In most springtails, gut contents could be observed in ordinary slides for species identification. The large, spherical species *Bourletiella hortensis* was squeezed on the slide to spread the gut content.

Figure 4. Sticky traps on a 6-year-old moraine, collecting airborne invertebrates from different directions, up to 1-m height. Inserted: open fallout trap with diameter 6.5 cm. From Ref. [28].

3. Succession patterns

3.1. Succession in species numbers

Figure 5 illustrates the cumulative species number for oribatid mites [24], springtails [25], and beetles and spiders [26], with increasing age of the ground. All groups showed a rapid addition of species during the first 80 years. Later, relatively few new species colonised among beetles and springtails. Oribatid mites and spiders, however, increased their cumulative species number considerably during the following 150 years. The five plots in 10,000-year-old soil had about the same number of springtail species as in 200-year-old soil, and with very few new species. Among oribatid mites, six new species were added. Beetles and spiders were not sampled in 10,000-year-old soil.

3.2. Succession in dominance structure

Another way of presenting succession is to examine the relative dominance among species. To illustrate the main changes among mites and springtails, data were lumped into the four mentioned age groups A–D (**Figure 2**), each with a similar number of sampling sites. In zone A (32–48 years), oribatids as a group made up about 80% of all mites, but this value later stabilised around 55%. The mite group Actinedida correspondingly increased their dominance in older soil, while predatory Gamasina mites were rare throughout the age gradient. In 10,000-year-old soil, the pioneer species *Tectocepheus velatus* was still present, but *Oppiella neerlandica* was now the dominant oribatid species [24].

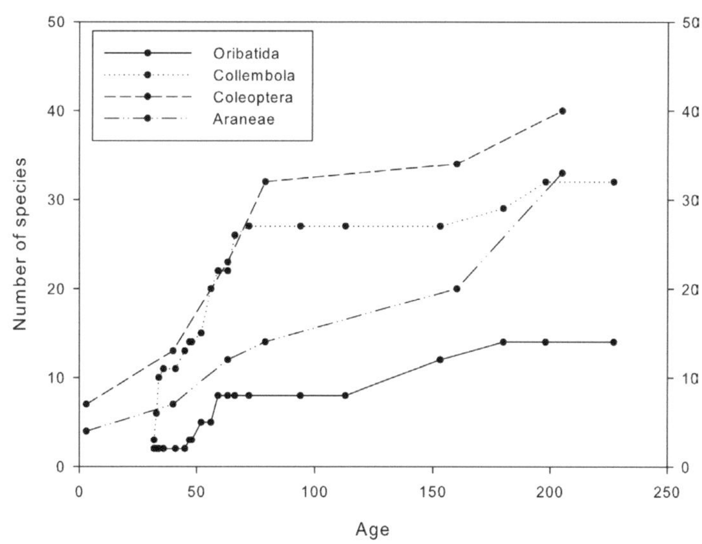

Figure 5. Cumulative species number of different arthropod groups, with increasing age of the ground. Surface active beetles (Coleoptera) and spiders (Araneae) were pitfall-trapped in various vegetation types, while springtails (Collembola) and oribatid mites (Oribatida) were extracted from soil in *Salix herbacea* vegetation. From Ref. [26].

Also, the springtail community showed considerable changes in dominance structure as the soil aged [25]. Two *Folsomia* species took over the dominance in zone B (52–66 years), and later *Tetracanthella brachyura* became the most abundant species. In short, there was a '*Folsomia* front' approaching the pioneer ground, and behind it followed a '*Tetracanthella* front'. The dominance structure of springtails was surprisingly similar in zone C (72–227 years) compared to the very old soil of 10,000 years in zone D [25].

Pitfall catches of beetles and spiders indicated clear changes in community structure during the 200-year study period. This is illustrated for beetles in **Table 1**, which shows the relative catches of the 13 most common species. *Bembidion hastii* dominated the catches strongly on 3-year-old ground but was still well represented on 40-year-old ground, for then to disappear on older ground. On 63- and 79-year-old ground, the community structure was very similar, being dominated by *Amara quenseli* and *Patrobus septentrionis*. While *A. quenseli* became very rare in older sites, *P. septentrionis* increased its dominance further and represented more than half of the catches on 160-year-old ground. However, on 205-year-old ground, *Liogluta alpestris* from the Staphylinidae family took over the dominance.

Table 2 lists the 14 most common spider species in the pitfall material. The catches on newly deglaciated ground was dominated by *Pardosa trailli, Erigone tirolensis* and *E. arctica*, while *Collinsia holmgreni* and *Hilaira* cf. *frigida* took over the dominance after 40 years. As in beetles,

Species	Family	3 yr	40 yr	63 yr	79 yr	160 yr	205 yr
Bembidion hastii	Carabidae	**81**	22	0	0	0	0
Nebria nivalis	Carabidae	**10**	**10**	1	2	<1	<1
Amara alpina	Carabidae	4	**23**	4	4	4	10
Geodromicus longipes	Staphylinidae	2	**10**	6	4	2	2
Simplocaria metallica	Byrrhidae	1	**9**	<1	<1	0	0
Amara quenseli	Carabidae	2	0	**30**	21	<1	1
Curimopsis cyclolepidia	Byrrhidae	0	0	**9**	3	0	0
Nebria rufescens	Carabidae	0	1	2	**9**	<1	<1
Patrobus septentrionis	Carabidae	0	23	29	32	**54**	13
Cymindis vaporariorum	Carabidae	0	0	5	1	**13**	1
Liogluta alpestris	Staphylinidae	0	0	7	11	6	**34**
Anthophagus alpinus	Staphylinidae	0	0	0	1	6	**13**
Chrysomela collaris	Chrysomelidae	0	0	0	<1	2	**8**
Other species		0	2	6	10	10	**16**
Total percentage		100	100	100	100	100	100

Table 1. Dominance structure of the beetle community on the ground of different age, expressed by pitfall catches. For each species, the highest dominance value is shown in bold numbers. Species with dominance values below 5% are collectively listed under 'Other species'.

http://dx.doi.org/10.5772/intechopen.68192

Species	Family	3 yr	40 yr	63 yr	79 yr	160 yr	205 yr
Pardosa trailli	Lycosidae	**36**	14	6	26	1	1
Erigone tirolensis	Linyphiidae	**27**	19	1	1	2	<1
Erigone arctica	Linyphiidae	**25**	2	16	18	0	<1
Collinsia holmgreni	Linyphiidae	12	**34**	1	2	3	1
Hilaira cf. *frigida*	Linyphiidae	0	**30**	5	10	7	6
Tiso aestivus	Linyphiidae	0	0	**34**	23	9	19
Arctosa alpigena	Lycosidae	0	0	**26**	17	11	0
Scotinotylus evansi	Linyphiidae	0	0	6	2	**26**	5
Pelecopsis mengei	Linyphiidae	0	0	3	0	**12**	<1
Pardosa septentrionalis	Lycosidae	0	0	0	0	**8**	0
Ozyptila arctica	Thomisidae	0	0	0	0	**8**	<1
Pardosa paludicola	Lycosidae	0	0	0	0	<1	**39**
Oedothorax retusus	Linyphiidae	0	0	0	0	0	**10**
Gonatium rubens	Linyphiidae	0	0	0	0	1	**8**
Other species		0	1	2	1	12	11
Total percentage		100	100	100	100	100	100

Table 2. Dominance structure of the spider community on the ground of different age, expressed by pitfall catches. For each species, the highest dominance value is shown in bold numbers. Species with dominance values below 5% are collectively listed under 'Other species'.

spiders showed a similar community structure on 63- and 79-year-old ground, being dominated by *Tiso aestivus, Arctosa alpigena* and *E. arctica*. After 160 years, *Scotinotylus evansi* dominated the catches, and *P. paludicola* dominated after 205 years.

3.3. Do species persist after colonisation?

A study of macroarthropod succession in several Norwegian glacier forelands at different altitudes and environmental conditions concluded that most species persisted after colonisation [19, 20, 22, 23]. This was regarded as a fundamental difference as compared to plant succession patterns. However, the taxonomic resolution in these studies was low in certain animal groups. For instance, in the beetle family Staphylinidae, which was represented by a high number of species in our study (21 out of 40 beetles), most species were unidentified in the studies by Vater and Matthews. The number of traps used in their studies was also low in some cases.

The more extensive material from the present case study, where all beetles and spiders were identified to species, confirmed to a large degree the hypothesis of 'adding and persistence' of species [26]. However, there were some exceptions. Among beetles, *B. hastii* disappeared when vegetation became more or less closed after about 60 years. *Simplocaria metallica* became very rare at the same time, and was not recorded after about 80 years. Likewise, the cold-adapted *Nebria nivalis* nearly disappeared after 80 years. *Curimopsis cyclolepidia* was only recorded in the range of 60–80 years. This range was also the clearly preferred for *A. quenseli*. It is interesting to note that the two last-mentioned species were not found in an extensive pitfall trapping in various neighbouring habitats of 10,000 years of age during 3 years (1969–1971) [36].

Among spiders, *P. trailli* and *E. arctica* nearly disappeared after 80 years. *A. alpigena* was numerous between 60 and 160 years but was absent after about 200 years. At 160 years, two new species appeared as very common: *P. septentrionalis* and *Ozyptila arctica*, but the first one disappeared in 200-year-old soil, and the second one nearly so.

Most soil microarthropods seemed to persist after colonisation. Among oribatid mites, the pioneer species *Liochthonius* cf. *sellnicki* was barely present after about 70 years [24]. Among springtails, the cold-loving pioneer species *Agrenia bidenticulata* gave up after about 50 years. There were several examples of rare species which 'disappeared' in older soils, but the small data do not allow firm conclusions about the presence or absence.

A general persistence of both macro- and microarthropods during succession indicates that these species have a high tolerance for each other and for changes in vegetation. Obviously, the concept of tolerance is as important as facilitation and inhibition when we try to understand succession.

3.4. Relations to environmental parameters

3.4.1. Parameters related to age

A detrended correspondence analysis (DCA) showed that terrain age was strongly correlated to the distance to the glacier, increased organic content in soil and falling pH values [24]. A species biplot of a DCA for mites sorted pioneer species, seral species and late seral species rather well into groups, confirming a successional process [24]. Correspondingly, a non-metric multidimensional scaling (NMDS) plot for springtails separated well the pioneer species. However, in contrast to the mite succession, which showed considerable difference between 72–227 years and 10,000 years, the NMDS plot for springtail communities confirmed a great similarity in these two age groups [25]. Concerning beetle and spider succession, an NMDS plot showed a clear succession, and the pioneer species were best separated. Also, vegetation cover was correlated with age and distance from the glacier [26].

Figure 6 shows a linear relation between age and thickness of the organic layer ($R = 0.83$, $F = 38.4$, $P < 0.001$). However, the variation was large. In **Figure 7**, species numbers of springtails and oribatid mites were related to the depth of the organic layer, and adapted curves indicated that species numbers tended to flatten out at about 10-mm organic layer. Pioneer microarthropod species have to be independent of an organic layer, and able to live on or close to the surface. Surface-living species are called epedaphic, litter-dwellers hemiedaphic and deeper-living species euedaphic. The two first groups typically contain larger species with eyes and pigmented body, while euedaphic species are often small, white and blind. **Figure 8** shows the number of springtail species from each of these categories (or transition

http://dx.doi.org/10.5772/intechopen.68192

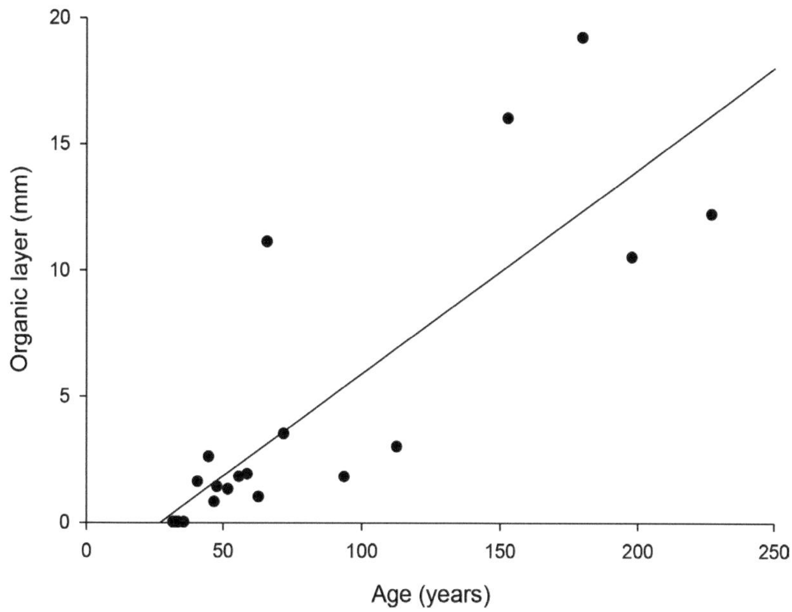

Figure 6. Relationship between the age of soil and the thickness of the organic layer.

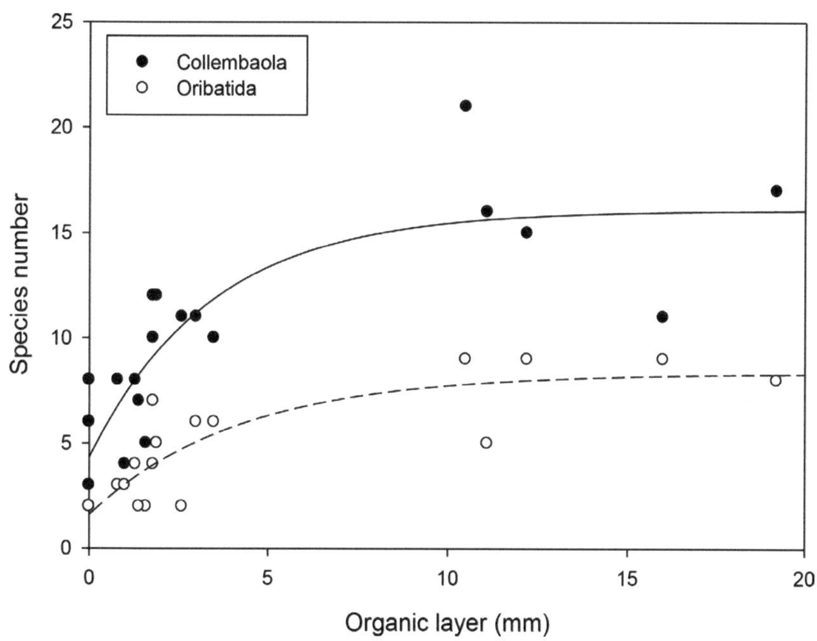

Figure 7. Adapted curves for the relationship between the thickness of the organic layer in soil and species numbers of springtails (Collembola) and oribatid mites (Oribatida).

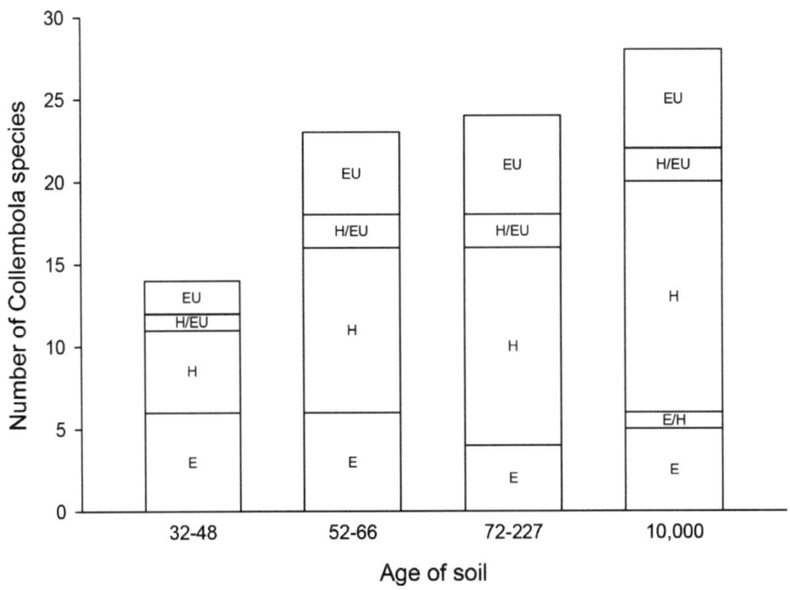

Figure 8. Structure of the springtail (Collembola) community at different age groups of the soil. E = Edaphic species, which are surface-living. H = Hemiedaphic species, which are litter dwellers. EU = Euedaphic species, which are deeper-living soil species.

categories) in soil of different age groups. It is a bit surprising that some hemiedaphic, and even two euedaphic species were recorded already in soil of 32–48 years of age. However, their presence may not be permanent. The organic layer was absent up to 36 years, and 1–3-mm thick in 41–48-year-old plots [24]. Later, the hemiedaphic species gradually became numerous, with 14 species in the very old soil. About five euedaphic species were established already at the age of 52–66 years, and the species number in this category changed little with further age.

3.4.2. A 'wet' and a 'dry' successional pathway

Local variation in soil moisture modified the succession pattern, among both surface-active macroarthropods and soil-living microarthropods. Direct correlation between soil moisture measured close to single traps, and the species collected there showed that the following beetles significantly preferred moist soil: *P. septentrionis, Geodromicus longipes* and *L. alpestris*. Three other species were clearly dry-ground dwellers: *Byrrhus fasciatus, Cymindis vaporariorum* and *A. quenseli* [37]. Among spiders, *T. aestivus* is an example of a dry-ground dweller, as all of 102 specimens were collected on dry ridges with lichen-dominated vegetation.

In **Figure 9**, the 'noise' from varying moisture conditions was identified by separating catches of beetles from typical 'wet' and typical 'dry' traps. We see how *P. septentrionis* dominated strongly in wet sites, while *A. quenseli* dominated the catches in dry sites nearby.

http://dx.doi.org/10.5772/intechopen.68192

Earlier studies have considered soil moisture to be the most important ecological factor for ground-living beetles in Norwegian alpine areas [38]. This is in accordance with our results. We conclude that surface-active macroarthropods followed two parallel successional trends in the foreland: a dry and a moist pathway.

Also, the succession of soil animals was affected by moisture. In the same glacier foreland, oribatid mites have been studied on dry moraine ridges of known age [18]. This allows for a comparison of the oribatid community in dry soil with neighbouring, moist snow bed soil at two age groups: 45–47 years and 66–72 years. While the generalist *T. velatus* was well represented in both dry and wet habitats, *L. lapponicus* and *Camisia horrida* occurred only on dry ridges, and *L.* cf. *sellnicki* and *C. foveolata* only in the wet snow bed. This illustrates that species within the same genus may have quite different moisture preferences.

Figure 10 illustrates schematically wet and dry succession. Both pathways were reflected in the vegetation mosaic of the foreland. This is in accordance with studies in the Rotmoos foreland in the Austrian Alps, where the effect of local topography and exposure on various invertebrate groups was studied. The moisture regime was an important factor on a local scale, for all site ages [7].

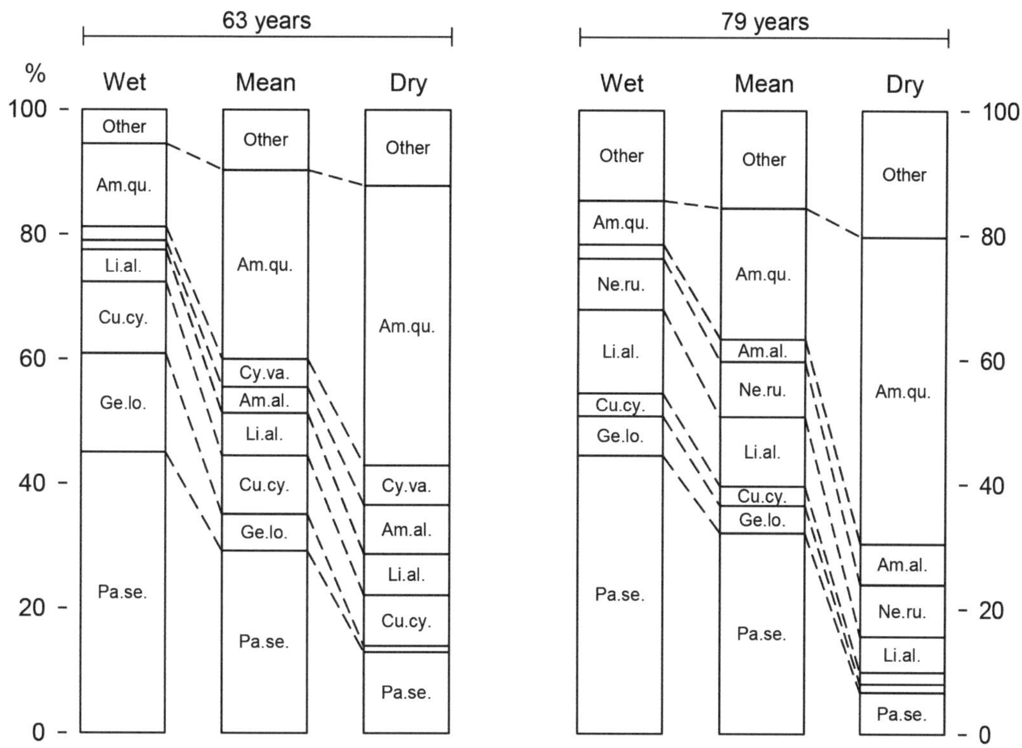

Figure 9. Effect of wet and dry ground on the structure of the beetle community. Both on 63- and 79-year-old plots, *Patrobus septentrionis* (Pa. se.) dominated on wet ground, and *Amara quenseli* (Am. qu.) on dry ground. Full names of the other species are *Amara alpina* (Am. al.), *Curimopsis cyclolepidia* (Cu. cy.), *Cymindis vaporariorum* (Cy. va.), *Geodromicus longipes* (Ge. lo.), *Liogluta alpestris* (Li. al.) and *Nebria rufescens* (Ne. ru.).

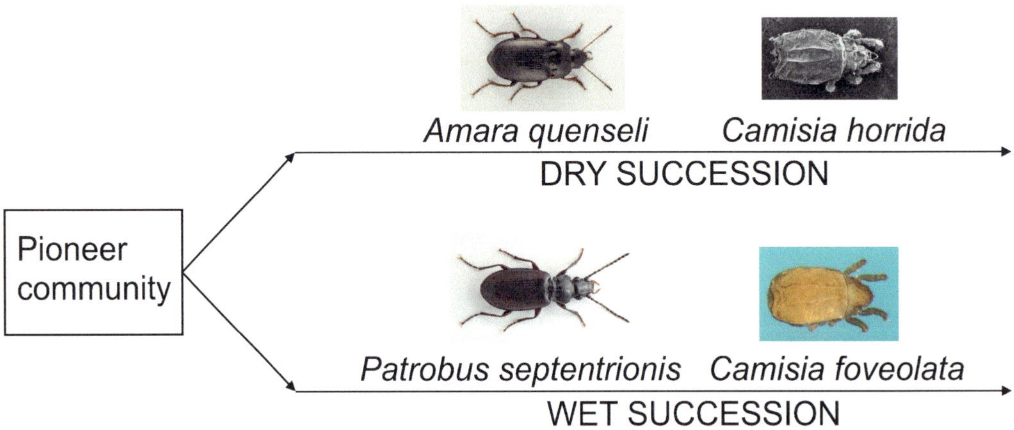

Figure 10. While the pioneer community is rather predictable, the further succession pattern differs in dry and wet patches. The figure shows a characteristic beetle and oribatid mite for a 'dry' and 'wet' succession, respectively.

3.5. Succession of surface animals versus soil animals

Among both surface-active macroarthropods and soil-living microarthropods, species numbers increased markedly during the first 80 years. Both groups had species that were favoured by the glacier retreat, because either they were cold-adapted (the springtail *A. bidenticulata* and the carabid beetle *N. nivalis*) or they preferred open space (the springtail *B. hortensis* and the carabid beetle *B. hastii*). Furthermore, both surface-living and soil-living animals were split into a 'dry' and a 'moist' succession pattern.

The two groups were, however, differently related to the development of vegetation. Soil-living microarthropods were favoured by the gradual development of an organic soil layer. Surface-active macroarthropods were influenced by the gradual closing of vegetation, and for some, to the appearance of food plants. While predators dominated throughout succession among macroarthropods, there were fewer predator species among microarthropods.

3.6. Comparison between plant and animal succession

Several investigators have detected a peak in plant richness early in primary succession, followed by a decline due to increased competition [4]. For instance, in the nearby foreland of Blåisen, there was an early diversity peak in proximal slopes [34]. Most arthropods, however, tend to persist after colonisation (see below), and there is no early peak in species richness.

Plant and animal succession have several features in common. Both plants and animals respond to local soil moisture, resulting in a 'dry' and a 'wet' succession. In the foreland of Blåisen, microtopography and moisture clearly affected plant succession [34].

Another similarity between botanical and zoological succession in the Finse area is that it takes at least 200 years to establish a stable "climax" community. Near Blåisen, only communities of simple structure, such as snow beds, reached a mature state after 220 years of succession [39].

Furthermore, plants with very narrow niches could attain local optima during early succession on glacier forelands and nunataks [35, 40]. Examples were *Draba cacuminum, Poa herjedalica, P. jemtlandica* and *Sagina intermedia* in the Finse area. Corresponding arthropod examples are two open space-living species: the carabid beetle *B. hastii* and the springtail *B. hortensis*, as well as two cold-loving species: the carabid beetle *N. nivalis* and the springtail *A. bidenticulata*.

A general similarity between plant and animal succession in Norwegian forelands is that the process is markedly affected by altitude and local climate. Glacier forelands in a harsh climate at high altitudes create a slow and species-poor succession, while the succession in both plants and animals is rapid and species-rich in forelands situated below the tree line [4, 19, 20, 22, 23].

3.7. Pioneer arthropods—a heterogenic group

The pioneer community was an interesting mix of generalists and specialists, and of various life strategies [26, 32]. Among early springtails and mites, there were both parthenogenetic and bisexual species, and species with either a short or a long life cycle [24, 25]. Furthermore, there were open-ground species as the springtail *B. hortensis* and the carabid beetle *B. hastii*, and 'cold-loving' species represented by the springtail *A. bidenticulata* and the carabid beetle *N. nivalis*. Several generalists colonised the pioneer ground. The harvestman *Mitopus morio* is a generalist predator, with high catches throughout the whole foreland [27]. Among oribatid mites, *T. velatus* is a well-known generalist, and among springtails we can point at *Desoria olivacea* and *Isotoma viridis*. The carabid *A. alpina* and the staphylinid *G. longipes* are habitat-tolerant beetles, and *E. tirolensis* is a spider example. Despite differences in ecology, pioneer arthropods have certain key abilities in common: they are good dispersers and can live, eat and reproduce on barren or nearly barren ground [32].

4. Dispersal: how to get there?

The rapid colonisation of newly exposed ground indicated that arthropods have a high dispersal ability. On Iceland, springtails and oribatid mites easily colonised recently emerged nunataks, and isolation of a few kilometres did not affect the colonisation [16]. These results strongly indicate aerial dispersal, and our study supports this. Fallout traps and sticky traps collected nine species of springtails and four species of oribatid mites, as well as some Actinedida mites and spiders (**Figure 11**). Among other items were unwinged aphids, some flies, several chironomid midges, a few seeds, and many fragments and diaspores of pioneer mosses [28]. Most aerial transport occurred rather close to the ground, below 0.5-m height. Sand grains in sticky traps up to this level illustrated the mechanical force of wind transport.

Some of the trapped species were assumed not to be able to thrive on pioneer ground, but to depend on a thicker organic layer [28]. In that case, their dispersal ability is high, but the pioneer ground may act as a 'sink' for them, where they will die. A 'real' pioneer species must

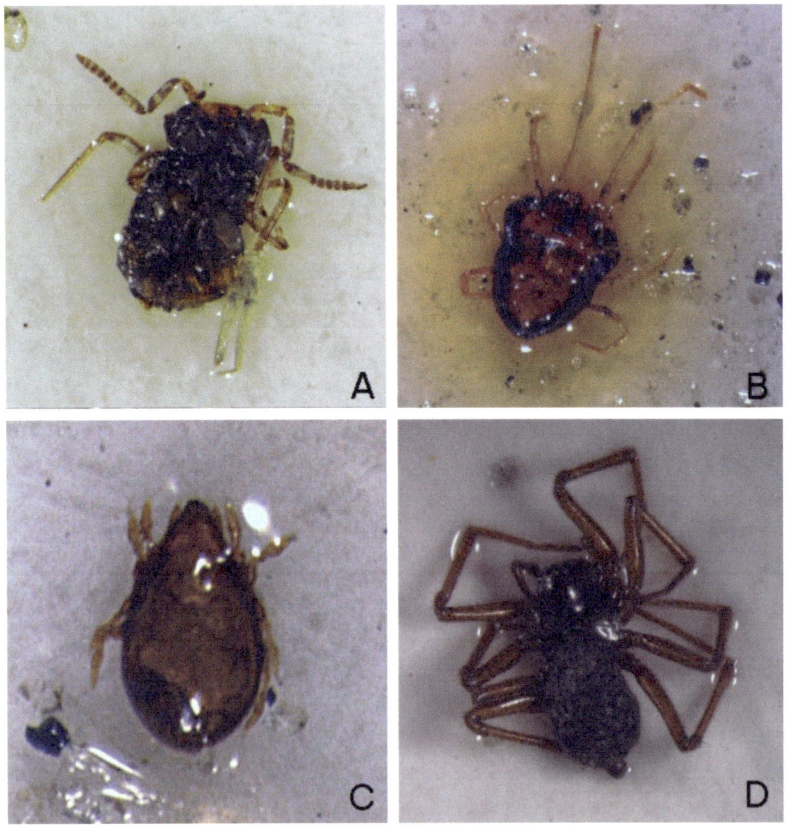

Figure 11. Invertebrates taken in sticky traps, proving airborne transport. A = the springtail *Bourletiella hortensis*. B = the mite *Bryobia* sp. C = the mite *Tectocepheus velatus*. D = the spider *Erigone arctica*. From Ref. [28].

be able both to arrive, to tolerate the harsh environmental conditions, to manage competition, to find food and to reproduce. In other words, pioneers must pass two 'filters': a 'dispersion filter' to arrive and an 'ecological filter' to establish a population.

5. Food sources: how to survive?

How can so many arthropod species—even predators—thrive on bare ground, before higher plants have established, or are very few? Based on analyses of the gut contents in springtails, beetles, harvestmen and chironomid midge larvae, we found that there were three 'invisible' food sources on newly deglaciated ground: biofilm with diatom algae, tiny pioneer mosses and ancient carbon delivered by the glacier.

5.1. Terrestrial biofilm as food

The springtail *A. bidenticulata* (**Figure 12**) was a 'super-pioneer', following the retreating glacier edge closely. Their guts contained a rather compact material dominated by tiny mineral particles, but diatom algae could often be seen [29, 31] (**Figure 13**). We assume that mineral

http://dx.doi.org/10.5772/intechopen.68192

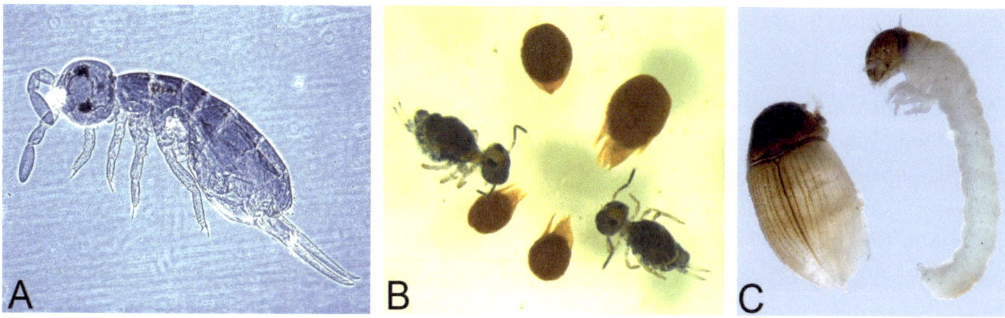

Figure 12. Some pioneer invertebrates. A = the biofilm-eating springtail *Agrenia bidenticulata*. B = the moss-eating springtail *Bourletiella hortensis*, together with four bulbils (dispersal units) of the moss *Pohlia* sp. C = newly hatched adult and a larva of the moss-eating beetle *Simplocaria metallica*. From Ref. [31].

particles were ingested accidentally when 'grazing' on biofilm. Terrestrial diatoms have the ability to establish a slimy, nutrient-rich biofilm on open ground by producing large quantities of extracellular polymeric substances [41, 42]. Diatom algae were also found in some guts of two other pioneer springtails: *D. olivacea* and *I. viridis* [31]. The early presence of terrestrial diatom algae shows that chlorophyll-based food chains start almost immediately after deglaciation.

5.2. Pioneer mosses as food

Already on a four-year-old ground, five mosses were observed: *Ceratodon purpureus, Bryum arcticum, Pohlia filum, Racomitrium canescens* and *Funaria hygrometrica* [31]. On nunataks of

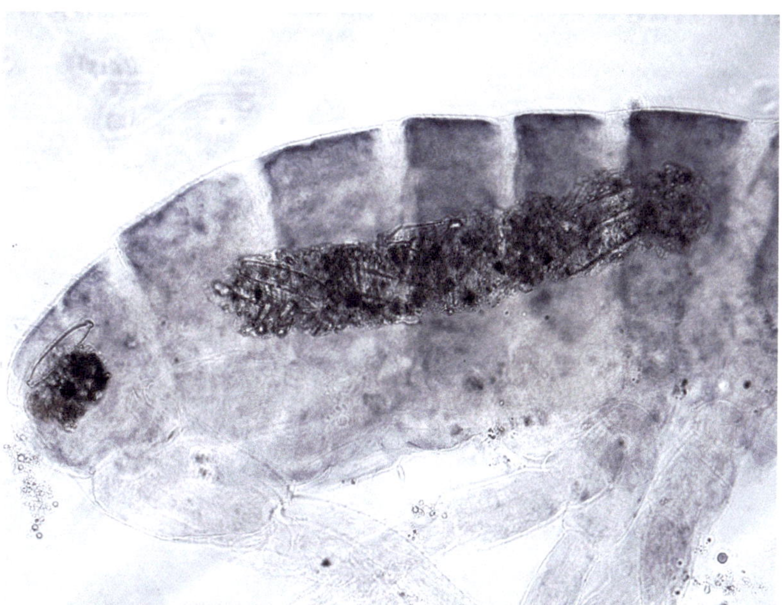

Figure 13. Hind part of the springtail *Agrenia bidenticulata* showing diatom algae in the gut. Most diatoms are densely packed, but a single one is easily seen to the left. From Ref. [31].

Omnsbreen glacier, about 10 km further North, a similar pioneer moss community has been found [40].

Due to characteristic cell structure in each moss species or genus, it was possible to identify small moss fragments in arthropod guts. On a 3-year-old ground, the relatively large and spherical springtail *B. hortensis* (**Figure 12**) had eaten leaves of *C. purpureus, Bryum* sp. and *Pohlia* sp., as well as nutrient-rich dispersal units (bulbils) of *P. filum* [31]. Among beetles, the family Byrrhidae is known to have moss-feeders, and on a six-year-old ground, guts of *S. metallica* (**Figure 12**) contained three different mosses (**Figure 14**). Two carabid beetles on three–six-year-old ground were omnivores, as their guts contained both invertebrates and moss fragments: *A. alpina* and *A. quenseli*. Conclusively, as much as four pioneer arthropods grazed on pioneer mosses [31].

5.3. Ancient carbon as food

The identification of this food resource was gradual, and surprising. A publication from a glacier foreland in the Austrian Alps showed that heterotrophic microbial communities used ancient carbon released by the glacier [43]. We wondered whether ancient carbon was released also by our glacier, and if so, whether it could be used as a nutrient source for pioneer arthropods. In September 2010, samples of surface soil (sand and silt) were taken 20 m from the glacier edge. During that summer, the glacier had retreated as much as 34 m. Analyses by Beta Analytic in Florida concluded that the samples contained material which was in average 21,000 years old. Furthermore, radiocarbon dating of chironomid midges and four predators, which were pitfall trapped on a 6–7-year-old ground showed that they all contained ancient carbon. The wolf spider *P. trailli*, had a radiocarbon age of 340 years, the harvestman *M. morio* 570 years, the carabid beetle *N. nivalis* 690 years, the carabid beetle *B. hastii* 1100 years and chironomid midges 1040 years [29]. Even larvae and adults of predatory diving beetles collected in young ponds had a radiocarbon age of 1100–1200 years. Springtails, however, did not contain ancient carbon.

New samples of surface soil taken close to the ice edge 4 years later corrected the age of released organic material to about 5160 years [30]. In the latter analysis, samples were pretreated at a lower temperature so that possible graphite particles from the phyllite-containing bedrock were not combusted and included in the analysis.

We found that chironomid larvae living in the sediment of young ponds assimilated the ancient carbon, and achieved a radiocarbon age up to 3270 years. We assumed that these larvae were eaten by diving beetles (**Figure 15**), and that adult midges ending on the soil

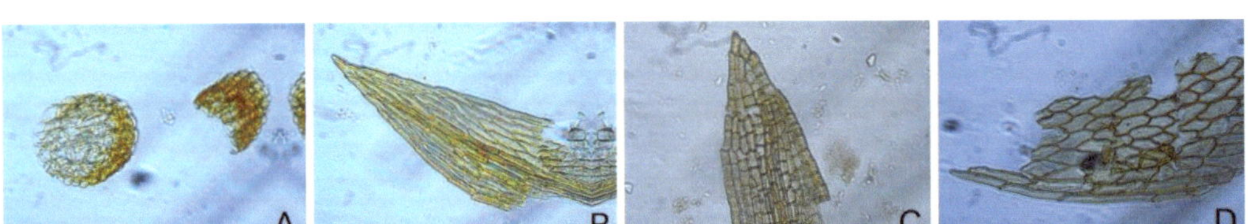

Figure 14. Moss fragments recorded in the gut of the beetle *Simplocaria metallica*. A = cross sections of a moss stem. B = leaf of *Pohlia* sp. C = leaf of *Ceratodon purpureus*. D = typical cell structure of a *Bryum* leaf. From Ref. [31].

surface after swarming fed terrestrial predators. Studies of the gut contents of the carabid beetles *N. nivalis* and *B. hastii*, and the harvestman *M. morio* confirmed that adult chironomid midges were an important part of their diet.

To be sure that ancient carbon was assimilated into the body tissue, measurements were also made on the larvae of Tipulidae (another Diptera group) in the same pond sediment, being careful to remove the gut contents before analysis. The actual body tissue from Tipulidae larvae had a radiocarbon age of 1610 years [30]. Moreover, chironomid larvae collected in the glacier river, and freed from their gut contents, had radiocarbon ages up to 1260 years.

We concluded that ancient organic material released by the glacier was assimilated by chironomid larvae, and transported further to aquatic and terrestrial predators. Chironomid midges thus supported early succession, and bound aquatic and terrestrial food webs together [29, 30].

The remaining question was: What is the source of the ancient carbon that had been stored in the glacier? We gradually abandoned the possibility that it came from old forest, bogs or soils from earlier periods where the glacier had been periodically absent. One reason was that the actual organic particles were probably extremely small. A purely chemical process, where carbon from non-biological bicarbonate served as a CO_2 source for aquatic algae, was also abandoned, since gut contents of chironomid larvae were practically free from algae [30]. Instead, our suspicion was led towards long-transported aerosols, originating from the incomplete combustion of fossil fuels. Such aerosols make up a part of the organic matter that glaciers collect by surface accumulation [30]. These aerosols are C_{14} depleted, and radiocarbon dating will reveal that they are very old [44, 45]. In fact, heavily glaciated watersheds may transport ancient, bioavailable carbon all the way to oceans, where marine microorganisms can assimilate the old carbon [46]. The aerosol hypothesis would fit with all our results [30].

5.4. A pioneer food web

Pioneer ground of 3–7 years of age contained a surprisingly diversity of food sources for pioneer arthropods (**Table 3**). Primary production was represented by invisible biofilm with

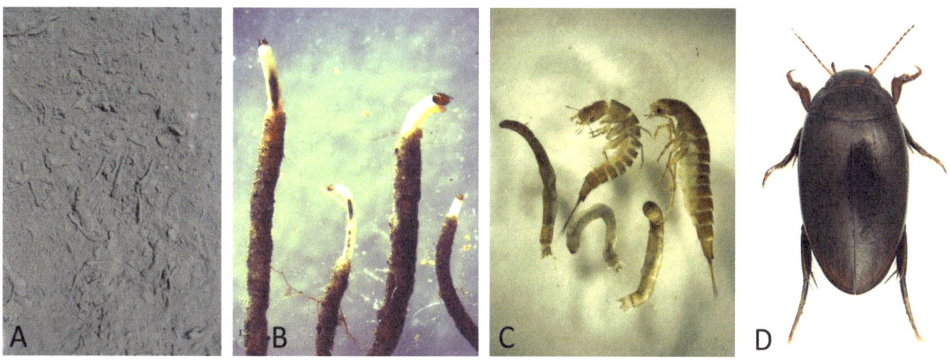

Figure 15. These pond-living invertebrates contained ancient carbon supplied by the melting glacier. A = sediment with chironomid larvae in tubes. B = chironomid larvae which have been partly freed from their tubes. C = two predacious larvae of the diving beetle *Agabus bipustulatus*, and three saprophagous, cylindrical larvae of Tipulidae (crane flies). D = adult predacious diving beetle, *Agabus bipustulatus*. From Ref. [30].

Species	Group	Biofilm	Fungal hyphae	Bryophytes	Vascular plants	Invertebrates	Ancient carbon via Chironomidae
Agrenia bidenticulata	Collembola	x					
Desoria olivacea	Collembola	x					
Isotoma viridis	Collembola	x	x				
Lepidocyrtus lignorum	Collembola		x				
Bourletiella hortensis	Collembola		x	x			
Simplocaria metallica	Coleoptera			x			
Amara alpina	Coleoptera			x	x	x	
Amara quenseli	Coleoptera			x		x	
Nebria nivalis	Coleoptera					x	x
Bembidion hastii	Coleoptera					x	x
Mitopus morio	Opiliones					x	x

Table 3. Food sources of terrestrial invertebrates on 3–6-year-old ground, based on gut content analyses. From Ref. [31].

diatom algae, tiny bryophytes and scattered vascular plants. Fungal hyphae found in some springtail guts were early terrestrial decomposers, and chironomids eaten by several predators were (as larvae) detritus feeders on ancient organic material. In addition, some inblown insects certainly contributed as prey. Two 'super-pioneers' followed the ice edge most closely: the biofilm-feeding springtails *A. bidenticulata* and *D. olivacea*.

To understand the food web on pioneer ground, we must combine aquatic and terrestrial food chains, and distinguish between chlorophyll-produced carbon, inblown carbon and ancient carbon released by the glacier. **Figure 16** summarises these relationships, and distinguishes between autotrophs, herbivores, predators and decomposers.

A pioneer food web can probably be of local character. In the present case, chironomid midges hatching from young ponds fed several terrestrial predators. In the Rotmoos foreland in Austria, however, springtails were found to be the main prey, and intraguild predation was demonstrated [47, 48].

5.5. Feeding categories during succession

Figure 17 shows that throughout the 200-year-old succession, the great majority of trapped macroarthropods were predators. While all spiders are predators, beetles contained a mixture of feeding categories. Pure herbivores were always represented by few species, even in the oldest sites.

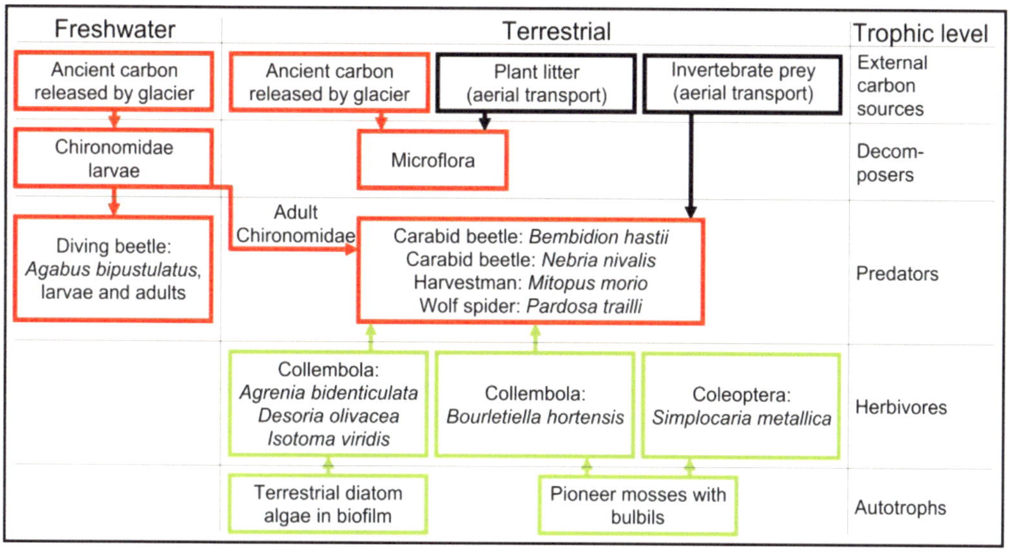

Figure 16. This food web from pioneer ground combines aquatic and terrestrial habitats. Shaded boxes illustrate the flow of ancient carbon, lower boxes with a grey frame show the flow of chlorophyll-produced carbon and the two upper boxes with a black frame show the use of carbon from inblown organic material. It is distinguished between autotrophs, herbivores, predators and decomposers. From Ref. [29].

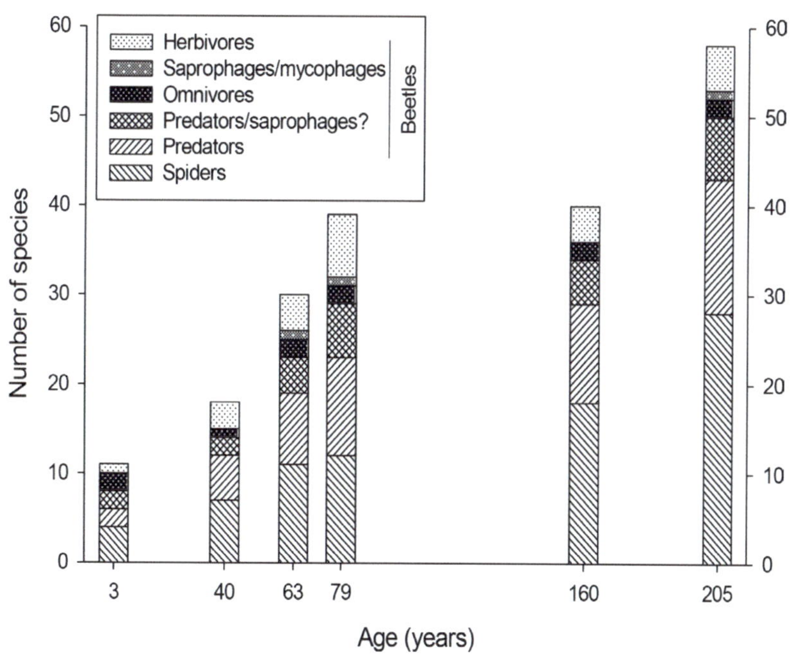

Figure 17. Feeding categories among pitfall-trapped macroarthropods at different ages of the ground. All spiders are predators, while beetles contained various feeding groups. Pure herbivores were rare throughout the gradient. From Ref. [26].

6. Driving forces in various phases of animal succession: facilitating and inhibiting factors

In early succession theory, facilitation, inhibition and tolerance were central concepts [49]. They were all used in a biotic context, and it was assumed that succession was driven by the way species interacted with one other. Early occupants could modify the environment in a way that influenced 'late-successional' species in three possible ways: (a) make the habitat more suitable for other species (facilitation) and (b) less suitable (inhibition), or early occupants had little or no effect on subsequent recruitment of species (tolerance). In the following presentation of four characteristic phases of succession, we use the terms facilitation, inhibition and tolerance in both biotic and abiotic contexts. We want to show that animal succession is only partly driven by the development of vegetation, and that abiotic factors may considerably influence the succession process.

6.1. Age 3–7 years: bare ground or only scattered pioneer vegetation

Wind facilitated transport of invertebrates, prey, algae and mosses into newly exposed ground [28]. In a foreland at Svalbard, aerial dispersal of midges and ballooning spiders was even assumed to add nutrients to virgin soil [12, 14].

The glacier itself facilitated the pioneer community by producing ponds, in which chironomid larvae assimilated ancient carbon. Within ponds, chironomid larvae were eaten by predatory diving beetles. Adult midges transported ancient carbon to terrestrial predators [29, 30]. The presence of predators before visible plants, often referred to as the 'predator first paradox' [13], can to a large degree be explained by local production of chironomid prey from young ponds. Cold-adapted species, like the springtail *A. bidenticulata* and the ground beetle *N. nivalis*, were facilitated by proximity to the glacier. However, a glacier retreat around 20 m per year means that they had to migrate continuously to remain in the cold zone.

6.2. Age about 30–40 years: patchy pioneer vegetation and much open ground

A high soil humidity due to much silt facilitated the colonisation of several plants and animals. Small patches of *S. herbacea* initiated the production of an organic layer. The moist-loving carabid beetle *P. septentrionis* colonised the ground. Pitfall catches documented a high surface activity among larger springtail species, not only within vegetated patches but also on bare ground [32].

6.3. Age about 60–200 years: mainly closed vegetation

A closed vegetation created shelter, reduced wind and maintained humidity. Web-building spiders were favoured by a three-dimensional vegetation. The pioneer ground beetle *B. hastii* disappeared when the vegetation became closed, but a local population survived on a 75-year-old bare patch [21]. The gradually deeper organic soil layer was positive for soil-living springtails and mites (**Figures 6–8**).

http://dx.doi.org/10.5772/intechopen.68192

For herbivores, the presence of a suitable food plant is crucial. While the moss-eating beetle *S. metallica* was found in the first moss patches on a 3-year-old ground, another moss-eating beetle, *B. fasciatus*, was not trapped until on a 63 years old ground. The beetle *Chrysomela collaris* feeds on *S. herbacea*, which occurs throughout the foreland. However, this beetle colonised late, and was found after 79 years. Clearly, other factors than the presence of the food plant determined the colonisation rate of some herbivorous beetles [26].

Both macro- and microarthropods were split into two main successional pathways: a dry and a wet succession. Due to patchy distribution of dry and moist habitats in the foreland, specialist on dry or moist sites had to overcome dispersal over unfavourable ground. The carabid beetle *C. vaporariorum*, which prefers dry ground, has a disadvantage by its inability to fly, due to rudimentary wings.

6.4. Age about 10,000 years: mature soil

The number of oribatid species increased clearly in this very old soil compared to 200-year-old soil, maybe facilitated by a deep organic layer. However, the increase was small for springtails, which were more efficient in colonising the foreland.

7. Remarks

Glacier forelands offer unique possibilities for the study of succession. We are beginning to understand patterns of arthropod succession by comparing studies from Norway, Svalbard, Iceland and the Alps [26, 32]. Several species or genera among arthropods are common pioneers in Norway and the Alps. However, glacier foreland chronosequences are both variable and complex. More case studies are needed, both to reveal local variations in pioneer communities and succession patterns and to look for general patterns.

Sample size is a critical factor. **Figure 18** shows how 12 soil samples within one plot gradually increased the cumulative number of mite taxa, but none of the samples contained all taxa. Ideally, sample numbers should be so high that the cumulative species number stabilises. If species numbers in different sites shall be compared, corresponding sampling effort should be used in all sites. To cover local variation in species composition, it may be better to take several small samples instead of a few larger covering the same area. During sampling with a soil corer, large, surface active springtails may escape by jumping. Some pitfall traps in addition may give valuable information.

Pitfall traps are much used for beetles and spiders in comparative studies, but the number of traps is often low. Even in the present study, with 20 traps operating during 2 years at each site, several species were taken in very few specimens [26]. Traps should be operated throughout the snow-free season, since certain species may have restricted seasonal activity.

The term 'primary succession' is questionable when both aquatic and terrestrial pioneer communities use ancient carbon released by the glacier. Young ponds acted as 'biological oases'

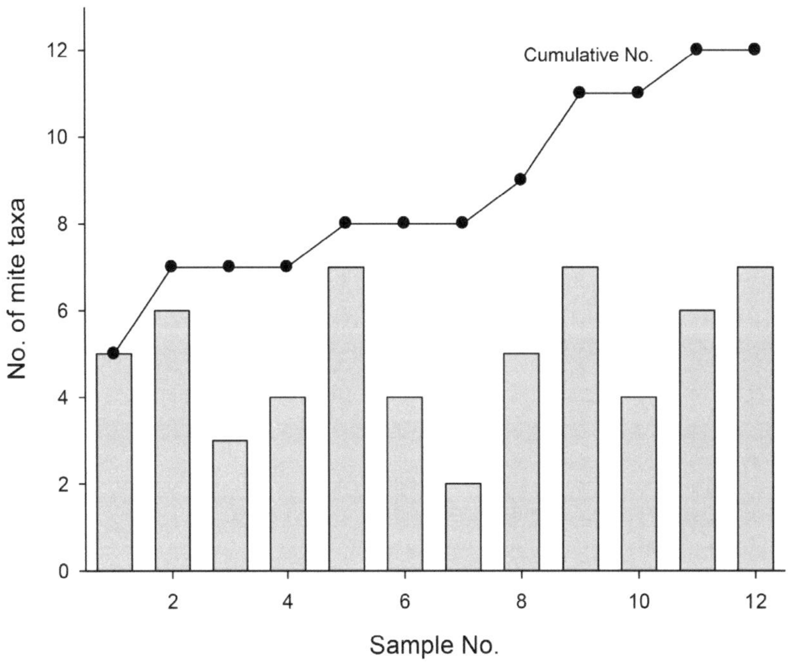

Figure 18. In plot no. 18 (age 180 years, see Figure 1), 12 soil cores were taken. This example shows how the cumulative number of mite taxa increased with increasing number of cores. However, none of the single cores contained all taxa (columns).

where ancient carbon was assimilated by Diptera larvae, mainly Chironomidae. In a foreland without ponds, a possible release of ancient carbon can be checked by radiocarbon dating chironomid larvae from the glacier river. A peculiar thing is that if invertebrates, which had assimilated old carbon, had been recovered as subfossils and radiocarbon dated, their age had been overestimated by up to 1100 years [29]. Since several pioneer species were herbivores on biofilm or mosses, the present succession did not fit with the 'predator-first' hypothesis. Although pioneer species may be ecologically very different, the pioneer community is surprisingly predictable, both within Norwegian forelands and in the Alps, and several genera are in common [32].

We need to improve our knowledge about the autecology of the individual species to better understand their position and functional role in the succession process. From each species′ point of view, colonising the foreland is a question of fulfilling minimum ecological demands. For instance, analyses of gut content were the key to understand the pioneer food web in the present foreland [29, 31]. Experimental studies involving transportation and re-location of species could be rewarding, but would it for the sake of science be ethically acceptable to move species within a 'natural laboratory' that should develop in a natural way?

A negative and special aspect by melting glaciers is that their meltdown will threaten cold-adapted invertebrates which live near glaciers. Especially when it comes to endemic, cold-adapted species, melting glaciers represent an extinction threat, as in certain mountains of the Southern Alps [50].

8. Conclusions

The questions posed in the 'Introduction' section can be answered in the following way:

a. *Comparison between botanical and zoological succession*

- Are there alternative succession patterns for arthropods, in the same way as there are alternative succession patterns for vegetation? Yes, also in animal succession, local conditions like topography and moisture modify the succession pattern. Among both soil-living microarthropods and surface-living macroarthropods, we can distinguish between a 'dry' and a 'moist' succession.
- Is there a strong progressive succession of arthropods on terrain ages of 20–50 years, as in plants? The period of fast colonisation among arthropods was longer in this study, around 80 years. Most species of springtails and beetles had arrived at that age, but species numbers of mites and spiders continued to increase substantially also after 80 years.
- Do animal and botanical successions differ in their early phases? Yes, in the early phase, animal succession may be said to differ from botanical succession. Before higher plants established, or were represented by very few species, a rather rich assemblage of arthropods was present. However, if we include pioneer mosses and terrestrial diatom algae, chlorophyll-based food chains established very early. In addition, several pioneer arthropods were able to use ancient carbon released by the glacier, being independent of primary production.

b. *Questions about zoological succession*

- Do most arthropod species tend to persist after colonising? Yes, but there were exceptions. The occurrence of many rare species makes it difficult to answer the question.
- Is a geo-ecological perspective fruitful and relevant when considering mechanisms of facilitation and inhibition in zoological successions? Yes, since abiotic factors highly influence colonisation and succession pattern.
- Does arthropod succession pattern differ between surface-active macroarthropods and soil-living microarthropods? There were large similarities.
- Is soil fauna succession in *S. herbacea* snow bed vegetation related to the gradual development of an organic layer? Yes, soil fauna succession goes on even after this vegetation type has been permanently established, and species numbers are related to the further development of the soil profile.

c. *Question about methods*

- Are sampling methods, material size and taxonomic resolution critical factors when studying arthropod succession? Yes.

Acknowledgments

We are grateful for being allowed to reuse figures and tables from Arctic, Antarctic and Alpine Research (copyright by the Regents of the University of Colorado) and Scientific Reports (copyright by Nature Publishing Group). We thank Ole Wiggo Røstad for producing several figures, Oddvar Hanssen for photographing beetles, Steve J. Coulson for the photograph of *C. foveolata* and Valerie Behan-Pelletier for SEM photograph of *Camisia horrida*.

Author details

Sigmund Hågvar[1]*, Mikael Ohlson[1] and Daniel Flø[2]

*Address all correspondence to: sigmund.hagvar@nmbu.no

1 Department of Ecology and Natural Resource Management, Norwegian University of Life Sciences, Ås, Norway

2 The Norwegian Institute of Bioeconomy Research, Ås, Norway

References

[1] Oerlemans J. Extracting a climate signal from 169 glacier records. Science. 2005;**308**:675-677.

[2] Jomelli V, Khodri M, Favier V, Brunstein D, Ledru M-P, Wagnon P, Blard P-H, Sicar, J-E, Braucher R, Grancher D, Bourlès DL, Braconnot P, Vuille M. Irregular tropical glacier retreat over the Holocene epoch driven by progressive warming. Nature. 2011;**474**:196-199.

[3] Malcomb NL, Wiles GC. Tree-ring-based reconstructions of North American glacier mass balance through the Little Ice Age – Contemporary warming transition. Quaternary Research. 2013;**79**:123-137.

[4] Matthews JA. The ecology of recently-deglaciated terrain: a geoecological approach to glacier forelands and primary succession. Cambridge: Cambridge University Press. 1992. 386 pp.

[5] Kaufmann R, Raffl C. Diversity in primary succession: the chronosequence of a glacier foreland. In: Körner C, Spehn E, editors. Global Mountain Biodiversity: A Global Assessment. London: Parthenon; 2002. pp. 177-190.

[6] Rydgren K, Halvorsen R, Töpper JP, Njøs JM. Glacier foreland succession and the fading effect of terrain age. Journal of Vegetation Science. 2014;**25**:1367-1380.

[7] Kaufmann R. Invertebrate succession on an Alpine glacier foreland. Ecology. 2001;**82**: 2261-2278.

[8] Kaufmann R, Fuchs M, Gosterxeier N. The soil fauna of an alpine glacier foreland: colonization and succession. Arctic, Antarctic and Alpine Research. 2002;**34**:242-250.

[9] Gobbi M, De Bernardi F, Pelfini M, Rossaro B, Brandmayr P. Epigean arthropod succession along a 154-year glacier foreland chronosequence in the Forni Valley (Central Italian Alps). Arctic, Antarctic, and Alpine Research. 2006;**38**:357-362.

[10] Gobbi M, Fontaneto D, De Bernardi F. Influence of climate changes on animal communities in space and time: the case of spider assemblages along an alpine glacier foreland. Global Change Biology. 2006;**12**:1985-1992.

[11] Gobbi M, Rossaro B, Vater A, De Bernardi F, Pelfini M, Brandmayr P. Environmental features influencing Carabid beetle (Coleoptera) assemblages along a recently deglaciated area in the Alpine region. Ecological Entomology. 2007;**32**:682-689.

[12] Hodkinson ID, Coulson SJ, Harrison J, Webb NR. What a wonderful web they weave: spiders, nutrient capture and early ecosystem development in the high Arctic – some counter-intuitive ideas on community assembly. Oikos. 2001;**95**:349-352.

[13] Hodkinson ID, Webb NR, Coulson SJ. Primary community assembly on land – the missing stages: why are the heterotrophic organisms always there first? Journal of Ecology. 2002;**90**:569-577.

[14] Coulson SJ, Hodkinson ID, Webb NR. Aerial dispersal of invertebrates over a high-Arctic glacier foreland: Midtre Lovénbreen, Svalbard. Polar Biology. 2003;**26**:530-537.

[15] Hodkinson ID, Coulson SJ, Webb NR. Invertebrate community assembly along proglacial chronosequences in the high Arctic. Journal of Animal Ecology. 2004;**73**:556-568.

[16] Ingimarsdóttir M, Caruso T, Ripa J, Magnúsdottir OB, Migliorini M, Hedlund K. Primary assembly of soil communities: disentangling the effect of dispersal and local environment. Oecologia. 2012;**170**(3):745-754. DOI: 10.1007/s00442-012-2334-8.

[17] Skubala P, Gulvik M. Pioneer oribatid mite communities (Acari, Oribatida) in newly exposed natural (glacier foreland) and anthropogenic (post-industrial dump) habitats. Polish Journal of Ecology. 2005;**53**:395-407.

[18] Seniczak A, Solhøy T, Seniczak S. Oribatid mites (Acari: Oribatida) in the glacier foreland at Hardangerjøkulen (Norway). Biological Letters. 2006;**43**:231-235.

[19] Vater AE. Invertebrate and arachnid succession on selected glacier forelands in southern Norway [thesis]. University of Wales; 2006. 472 pp.

[20] Vater AE. Insect and arachnid colonization on the Storbreen glacier foreland, Jotunheimen, Norway: persistence of taxa suggests an alternative model of succession. The Holocene. 2012;**22**(10):1123-1133.

[21] Alfredsen AN. Primary succession, habitat preferences and species assemblages of carabid beetles in front of the retreating glacier Midtdalsbreen, Finse, southern Norway [thesis]. University of Bergen; 2010. 83 pp.

[22] Vater AE, Matthews JA. Testing the "addition and persistence" model of invertebrate succession in a subalpine glacier-foreland chronosequence: Fåbergstølsbreen, southern Norway. The Holocene. 2013;**23**:1151-1162.

[23] Vater AE, Matthews JA. Succession of pitfall-trapped insects and arachnids on eight Norwegian glacier forelands along an altitudinal gradient: patterns and models. The Holocene. 2015;**25**(1):108-129.

[24] Hågvar S, Solhøy T, Mong C. Primary succession of soil mites (Acari) in a Norwegian glacier foreland, with emphasis on Oribatid species. Arctic, Antarctic, and Alpine Research. 2009;**41**:219-227.

[25] Hågvar S. Primary succession of springtails (Collembola) in a Norwegian glacier foreland. Arctic, Antarctic, and Alpine Research. 2010;**42**:422-429.

[26] Bråten AT, Flø D, Hågvar S, Hanssen O, Mong CE, Aakra K. Primary succession of surface active beetles and spiders in an alpine glacier foreland, central south Norway. Arctic, Antarctic, and Alpine Research. 2012;**44**(1):2-15.

[27] Hågvar S, Flø D. Succession and phenology of the generalist predator *Mitopus morio* (Fabricius, 1799) (Opiliones) in a glacier foreland. Norwegian Journal of Entomology. 2015;**62**:210-215.

[28] Flø D, Hågvar S. Aerial dispersal of invertebrates and mosses close to a receding alpine glacier in Southern Norway. Arctic, Antarctic, and Alpine Research. 2013;**45**:481-490.

[29] Hågvar S, Ohlson M. Ancient carbon from a melting glacier gives high ^{14}C age in living pioneer invertebrates. Scientific Reports. 2013;**3**:2820. DOI: 10.1038/srep02820.

[30] Hågvar S, Ohlson M, Brittain JE. A melting glacier feeds aquatic and terrestrial invertebrates with ancient carbon and supports early succession. Arctic, Antarctic, and Alpine Research. 2016;**48**:551-562.

[31] Hågvar S, Pedersen A. Food choice of invertebrates during early glacier foreland succession. Arctic, Antarctic, and Alpine Research. 2015;**47**:561-572. DOI: http://dx.doi.org/10.1657/AAAR0014-046

[32] Hågvar S. Primary succession in glacier forelands: how small animals conquer new land around melting glaciers. In: Young SS, Silvern SE, editors, International Perspectives on Global Environmental Change. INTECH Open Access Publisher; 2012. pp. 151-172. Free online on www.intechopen.com

[33] Matthews JA, Whittaker RJ. Vegetation succession on the Storbreen glacier foreland, Jotunheimen, Norway: a review. Arctic and Alpine Research. 1987;**19**:385-395.

[34] Elven R, Ryvarden L. Dispersal and primary establishment of vegetation. In: Wielgolaski FE, editor. *Fennoscandian Tundra Ecosystems, Part I*. Berlin: Springer Verlag; 1975. pp. 81-85.

[35] Elven R. Association analysis of moraine vegetation at the glacier Hardangerjøkulen, Finse, South Norway. Norwegian Journal of Botany. 1978;**25**:171-191.

[36] Østbye E, Hågvar S. Pit-fall catches of surface-active arthropods in high mountain habitats at Finse, south Norway. IV. Coleoptera. Fauna Norvegica Series B. 1996;**43**:1-18.

[37] Bråten AT, Flø D. Primary succession of arthropods (Coleoptera and Araneae) on a newly exposed glacier foreland at Finse, southern Norway [thesis]. Norwegian University of Life Sciences; 2009. 85 pp.

[38] Ottesen PS. Niche segregation of terrestrial alpine beetles (Coleoptera) in relation to environmental gradients and phenology. Journal of Biogeography. 1996;**23**:353-369.

[39] Elven R. Plant communities on recently deglaciated moraines at Finse, southern Norway. In: Vik R, editor. IBP in Norway: Methods and Results. Sections PT-UM Grazing Project, Hardangervidda, Botanical Investigations. Annual Report 1974. Appendix I. Oslo, Norwegian National IBP Committee; 1975. pp. 381-467.

[40] Elven R. The Omnsbreen glacier nunataks – a case study of plant immigration. Norwegian Journal of Botany. 1980;**27**:1-16.

[41] Heyl K, Woelfel J, Schumann R, Karsten U. Microbial mats from wind flats of the southern Baltic Sea. In: Seckbach J, Oren A, editors, Microbial mats. Modern and ancient microorganisms in stratified systems. Springer; 2010. pp. 303-319.

[42] Underwood GJC. Exopolymers (Extracellular polymeric substances) in diatom-dominated marine sediment biofilms. In: Seckbach J, Oren A, editors, Microbial mats. Modern and ancient microorganisms in stratified systems. Springer; 2010. pp. 289-300.

[43] Bardgett RD, Richter A, Bol R, Garnett MH, Bäumler R, Xu X, Lopez-Capel E, Manning DAC, Hobbs PJ, Hartley IR, Wanek W. Heterotrophic microbial communities use ancient carbon following glacial retreat. Biology Letters. 2007;**3**:487-490.

[44] Stubbins A, Hood E, Raymond PA, Aiken GR, Sleighter RL, Hernes PJ, Butman D et al. Anthropogenic aerosols as a source of ancient dissolved organic matter in glaciers. Nature Geoscience. 2012;**5**:198-201.

[45] Fellman JB, Hood E, Raymond PA, Stubbins A, Spencer RGM. Spatial variation in the origin of dissolved organic carbon in snow on the Juneau icefield, Southeast Alaska. Environmental Science and Technology. 2015;**49**:11492-11499. DOI: 10.1021/acs.est.5b02685

[46] Hood E, Fellman J, Spencer RGM, Hernes PJ, Edwards R, D'Amore D, Scott D. Glaciers as a source of ancient and labile organic matter to the marine environment. Nature. 2009;**462**:1044-1048.

[47] König T, Kaufmann R, Scheu S. The formation of terrestrial food webs in glacier foreland: evidence for the pivotal role of decomposer prey and intraguild predation. Pedobiologia. 2011;**54**:147-152.

[48] Raso L, Sint D, Mayer R, Plangg S, Recheis T, Brunner S, Kaufmann R, Traugott M. Intraguild predation in pioneer predator communities of alpine glacier forelands. Molecular Ecology. 2014;**23**:3744-3754.

[49] Connell JH, Slatyer RO. Mechanisms of succession in natural communities and their role in community stability and organization. American Naturalist. 1977;**111**:1119-1144.

[50] Tampucci D, Gobbi M, Boracchi P, Cabrini E, Compostella C, Mangili F, Marano G, Pantini P, Caccianiga M. Plant and arthropod colonisation of a glacier foreland in a peripheral mountain range. Biodiversity. 2015;**16**:213-223.